Andrew Lang

The Red Book of Animal Stories

Andrew Lang

The Red Book of Animal Stories

ISBN/EAN: 9783337006969

Printed in Europe, USA, Canada, Australia, Japan

Cover: Foto ©berggeist007 / pixelio.de

More available books at **www.hansebooks.com**

THE RED BOOK
OF
ANIMAL STORIES

SELECTED AND EDITED
BY
ANDREW LANG

WITH NUMEROUS ILLUSTRATIONS BY H. J. FORD

LONGMANS, GREEN, AND CO.
39 PATERNOSTER ROW, LONDON
NEW YORK AND BOMBAY
1899

TO

MISS SYBIL CORBET

*Author of 'Animal Land' 'Sybil's Garden of Pleasant Beasts'
and 'Epiotic Poems'*

Sybil, the Beasts we bring to you
  Are not so friendly, not so odd,
As those that all amazed we view,
  The brutes created by your nod—
  The Wuss, the Azorkon, and the Pod;
But then *our* tales are true!

Fauna of fancy, one and all
  Obey your happy voice, we know;
A garden zoological
  Is all around, where'er you go.
  Mellys and Kanks walk to and fro,
And Dids attend your call.

We have but common wolves and bears,
  Lion and leopard, hawk and hind,
Tigers, and crocodiles, and hares:
  But yet they hope you will be kind,
  And mark with sympathetic mind
These moving tales of theirs.

PREFACE

CHILDREN who read this book will perhaps ask whether all the stories are true? Now all the stories are *not* true; at least, we never meet the Phœnix now in any known part of the world. To be sure, there are other creatures, such as the Mastodon and the Pterodactyl, which are not found alive anywhere, but their bones remain, turned into stones or fossils. It is unlikely that they were changed into rocks by a witch, or by Perseus with the Gorgon's Head, in the Greek story. It must have been done in some other way. However, the bones, now stones, show that there were plenty of queer beasts that have died out. Possibly the sight of the stone beasts and birds made people believe, long ago, in such creatures as Dragons, and the water-bulls that haunt the lochs in the Highlands. One of these was seen by a shepherd about eighty years since, and an account of it was sent to Sir Walter Scott. There is also the Bunyip, a strange creature which both white and black men say that they have seen in the lakes of Australia. Then there is the Sea Serpent; many people have seen him alive, but no specimen of a dead Sea Serpent is in any of the museums. About 1,300 years ago, more or less, St. Columba saw a great water-beast, which lived in the river Ness, and roared as it pursued men; but the Saint put an end to its adventures. For my part, I do not disbelieve that there may be plenty of strange animals which scientific men have not yet dissected and named by long names. Some of the last of these may have been remembered and called Dragons. For, if there were never any Dragons, why did all sorts of nations tell stories about them? The Fire Drake, however, also the Ice Beast, or Remora, do seem very unlikely creatures, and perhaps they are only a sort of poetical inventions. The stories about these unscientific animals are told by Mr. H. S. C. Everard, who found them in very curious old books.

The stories about Foxes are by Miss B. Grieve, who is a great friend of Foxes, and takes their side when they are hunted by the Duke of Buccleuch's hounds. I am afraid she would not tell where the Fox was hiding, if she knew (as she sometimes does), just as you would not have told his enemies, if you had known that Charles II. was hiding in the oak tree. Not that it is wrong to hunt foxes, but a person who is not hunting naturally takes the weaker side. And, after all, the fun is to pursue the fox, not to catch him. The same lady wrote about sheep in 'Sheep Farming on the Border.'

The stories about 'Tom the Bear' are taken from the French works on natural history by M. Alexandre Dumas. We cannot be sure that every word of them is true, for M. Dumas wrote novels chiefly, which you must read when you are older. One of these novels is about Charles I., and it is certainly not all true, so we cannot believe every word that M. Dumas tells us. He had a great deal of imagination—enough for about thirteen thousand living novelists.

Most of the other tales are written by Mrs. Lang, and are as true as possible; while Miss Lang took the adventures of a Lion Tamer, and 'A Boar Hunt by Moonlight,' out of French and German books. The story of greedy Squouncer, by Mrs. Lang, is true, every word, and I wrote 'The Life and Death of Pincher,' who belonged to a friend of mine.[1] Squouncer's portrait is from a photograph, and does justice to his noble expression.

Miss Blackley also did some of the stories. Most of the tales of 'Thieving Dogs and Horses' were published, about 1819, by Sir Walter Scott, in 'Blackwood's Magazine,' from which they are taken by Mrs. Lang.

I have tried to make it clear that this is not altogether a *scientific* book; but a great deal of it is more to be depended on than 'A Bad Boy's Book of Beasts,' or Miss Sybil Corbet's books, 'Animal Land,' and 'Sybil's Garden of Pleasant Beasts.'

[1] From *Longman's Magazine*.

These are amusing, but it is not true that 'the Garret Lion ate Sybil's mummy.' Indeed, I think that when people, long ago, invented the Fire Drake, and the Ice Beast, they were just like Miss Corbet, when *she* invented the Kank, the Wuss, and other animals. That is to say, they were children in their minds, though grown up in their bodies. They fancied that they saw creatures which were never created.

If this book has any moral at all, it is to be kind to all sorts and conditions of animals—that will let you. Most girls are ready to do this, but boys used to be apt to be unkind to Cats when I was a boy. There is no reason why an exception should be made as to Cats, and a boy ought to think of this before he throws stones or sets dogs at a cat. Now, in London, we often see the little street boys making friends with every cat they meet, but this is not so common in the country. If anything in this book amuses a boy, let him be kind to poor puss, and protect her, for the sake of his obedient friend,

ANDREW LANG.

CONTENTS

	PAGE
The Phœnix	1
Griffins and Unicorns . . .	4
About Ants, Amphisbænas, and Basilisks .	12
Dragons	20
The Story of Beowulf, Grendel, and Grendel's Mother .	33
The Story of Beowulf and the Fire Drake . .	43
A Fox Tale	49
An Egyptian Snake Charmer	55
An Adventure of Gérard, the Lion Hunter .	61
Pumas and Jaguars in South America . . .	84
Mathurin and Mathurine . . .	98
Joseph: Whose proper name was Josephine .	102
The Homes of the Vizcachas .	108
Guanacos: Living and Dying .	112
In the American Desert .	117
The Story of Jacko II. . .	128
'Princess' . . .	135
The Lion and the Saint	138
The Further Adventures of 'Tom,' a Bear, in Paris .	143
Recollections of a Lion Tamer . .	154
Sheep Farming on the Border . .	171
When the World was Young . . .	177
Bats and Vampires . . .	196
The Ugliest Beast in the World	200
The Games of Orang-Outangs, and Kees the Baboon .	206
Greyhounds and their Masters . . .	224

	PAGE
The Great Father, and Snakes' Ways	232
Elephant Shooting	238
Hyenas and Children	252
A Fight with a Hippopotamus	257
Kanny, the Kangaroo	261
Collies, or Sheep Dogs	266
Two Big Dogs and a Little One	273
Crocodile Stories	280
Lion-Hunting and Lions	285
On the Trail of a Man-eater	304
Greyhounds and their Arab Masters	310
The Life and Death of Pincher	317
A Boar Hunt by Moonlight	321
Thieving Dogs and Horses	328
To the Memory of Squouncer	339
How Tom the Bear was born a Frenchman	344
Charley	357
Fairy Rings; and the Fairies who make them	364
How the Reindeer Live	370
The Cow and the Crocodile	376

ILLUSTRATIONS

PLATES

The Lion falls in love with Aïssa	Frontispiece	
The Griffin	to face p.	4
How the Unicorn was Trapped . .	,,	8
Finding a Mermaid	,,	16
Victor carried up the Chasm by the Dragon . .	,,	26
Queen Waltheow and Beowulf	,,	34
Grendel's Mother drags Beowulf to the bottom of the Lake	,,	38
The Death of Beowulf	,,	44
The Lion falls in love with Aïssa . . .	,,	62
Aïssa's Father finds her Axe	,,	70
The Lion appears at the top of the Ravine . .	,,	78
Maldonada guarded by the Puma . . .	,,	88
The Jaguar besieged by Peccaries	,,	92
Joseph's Breakfast	,,	104
St. Jerome draws out the Thorn . .	,,	138
Tom frightens the Little Girl .	,,	144
Just in time to save Tom	,,	150
Securing a Mammoth	,,	178
Megatheria . .	,,	184
The Vampire Bat	,,	196
How the Namaquas hunt the Rhinoceros . .	,,	202
Orang-Outangs eating Oysters on the Sea-shore .	,,	208
The Orang determines to throw the rival Monkeys overboard	,,	212

	PAGE
When this Prize was laid at the feet of the Lady, the Giver might ask in return for anything he chose . . . . to face p. 224	
Baker shooting the Elephants at the Island . . „ 240	
Hannibal's Elephants . . . . . . „ 248	
The Lion was in the air close to him . . . „ 290	
The Woodman and the Lions get the best of the Bear . . . . „ 296	
The Highwayman's Horse . . . . . „ 334	
The Captain had a Strange Dream . . . „ 346	
The Bear instantly rose on its hind legs and began to Dance . . . . „ 352	
Then a soft nose touched him . . . . „ 358	

IN TEXT

The Phœnix	2
The Odenthos	13
The Demon of Cathay . . .	15
Ragnar does battle with the Serpents . .	23
De Gozon and his Dogs fight the Dragon . .	31
The Snake Charmer	57
The Lion said to the Gazelles, 'Do not flee' . .	67
The Lion laughs at the Marabout's Question . .	75
Mathurin and Mathurine . . .	99
Spaniards meeting a Caravan of Llamas . .	113
Watching the Combat . . .	121
The Moccason Snake fascinates the Orioles	123
'Princess' and the Invalid	136
The Lion rescues the Ass from the Caravan .	142
I seized him by the scruff of the neck . . .	159
The Lion Tamer offers to wake the (stuffed) Crocodile! .	163
Digging the imprisoned Sheep out of the Snow . .	175
Stegosaurus	189

ILLUSTRATIONS

	PAGE
Pterodactyl	193
Le Vaillant and Kees out hunting	217
The Baboon who looked after the Goats	221
The Snakes found in the Lame Man's Bed	235
Oswell's narrow Escape	245
How the Hippopotamus attacked the Boat	259
The New Arrival	262
Kanny frightens the Carpenters	264
The Faithful Messenger	267
Finding the Necklace	283
The Lion in the Camp	301
Cumming's Cap frightens the Tiger	305
The Elephant tried to gore the Tiger with his Tusks	308
The Summons to the Hunt	313
Vomhammel in Danger	325
A Portrait of Greedy Squouncer	341
Hunting the Bison	367

THE PHŒNIX

In former times, when hardly anybody thought of travelling for pleasure, and there were no Zoological Gardens to teach us what foreign animals and birds were really like, men used to tell each other stories about all sorts of strange creatures that lived in distant lands. Sometimes these tales were brought by the travellers themselves, who loved to excite the wonder of their friends at home, and knew there was nobody to contradict them. Sometimes they may have been invented by people to amuse their children; but, anyway, the old books are full of descriptions of birds and beasts very interesting to read about.

One of the most famous of these was the Phœnix, a bird whose plumage was, according to one writer, 'partly red and partly golden,' while its size was 'almost exactly that of the eagle.' Once in five hundred years it 'comes out of Arabia,' says one old writer, 'all the way to Egypt, bringing the parent bird, plastered over with myrrh, to the Temple of the Sun (in the city of Heliopolis), and then buries the body. In order to bring the body, they say, it first forms a ball of myrrh as big as it can carry, puts the parent inside, and covers the opening with fresh myrrh; the ball is then exactly the same weight as at first; thus it brings the body to Egypt, plastered over as I have said, and deposits it in the Temple of the Sun.' This is all that the writer we have been quoting seems to know about the Phœnix; but we are told by someone else that its song was 'more beautiful than that of any other bird,' and that it was 'a very king of the feathered tribes, who followed it in fear, while it flew swiftly along, rejoicing as a bull in its strength.' Flashing its brilliant plumage in the sun, it went its way till it

THE PHŒNIX

reached the town of Heliopolis. 'In that city,' says another writer, whose account is not quite the same as the story told by the first—'in that city there is a temple made round, after the shape of the Temple at Jerusalem. The priests of that temple date their writings from the visits of the Phœnix, of which there is but one in all the world. And he cometh to burn himself upon the altar of the temple at the end of five hundred years, for so long he liveth. At the end of that time the priests dress up their altar, and put upon it spices and sulphur, and other things that burn easily. Then the bird Phœnix cometh and burneth himself to ashes. And the first day after men find in the ashes a worm, and on the second day they find a bird, alive and perfect, and on the third day the bird flieth away. He hath a crest of feathers upon his head larger than the peacock hath, his neck is yellow and his beak is blue; his wings are of purple colours, and his tail yellow and red in stripes across. A fair bird he is to look upon when you see him against the sun, for he shineth full gloriously and nobly.'

It is very hard to believe that the man who wrote this had not actually seen this beautiful creature, he seems to know it so well, and perhaps sometimes he really fancied that one day it had dazzled his eyes as it darted by. The Phœnix was a living bird to old travellers and those to whom they told their stories, although they are not quite agreed about its habits, or even about the manner of its death. Sometimes, as we have seen, the Phœnix has a father, sometimes there is only one bird. In general it burns itself on a spice-covered altar; but, according to one writer, when its five hundred years of life are over it dashes itself on the ground, and from its blood a new bird is born. At first it is small and helpless, like any other young thing; but soon its wings begin to show, and in a few days they are strong enough to carry the parent to the city of Heliopolis, where, at sunrise, it dies. The new Phœnix then flies back home, where it builds a nest of sweet spices—cassia, spikenard and cinnamon; and the food that it loves is another spice, drops of frankincense.

GRIFFINS AND UNICORNS

Some of the creatures that we read about in the books of the old travellers are quite easy to believe in, for, after all, they are not very unlike the birds and beasts that are to be seen to-day in different parts of the world. The Phœnix, though bigger, was not more beautiful than the tiny humming birds that dart through tropical forests, nor more splendid than the noisy macaws, and we can picture it to ourselves without any difficulty. But nobody now will ever go in search of the gourd that grows on a tree, and contains a little flesh-and-blood lamb; or expect, in travelling through Scotland, to find a Barnacle-Goose tree, with ducks instead of fruit, as a very clever gentleman who later became Pope did about four hundred and fifty years ago!

To us, who can look at a giraffe or a rhinoceros any day we choose, there is nothing so particularly strange about a griffin, which had the body of a lion, and the wings and head of an eagle, and was as strong as ten lions, or a hundred eagles. 'He will carry,' we are told, 'flying to his nest, a great horse, or two oxen yoked together as they go at the plough, or a man in full armour. For he hath his talons (claws) so long and so large and great upon his feet, as though they were the horns of great oxen, so that men make cups of them to drink of: and of his ribs and wing-feathers they make a very strong bow, to shoot with arrows and querrels.' A 'querrel,' it is needful to explain, was a bolt shot from a crossbow.

The Griffin

Griffins were not to be met with every day, nor in every country; but they roamed freely through the Caucasus Mountains, in search of gold and precious stones. Indeed, so fond of gold was the griffin, that after he had dug out a large heap with his powerful claws, he would roll about in it with delight, or sit and look at it by the hour together.

But, unluckily, the griffin was not allowed to enjoy this innocent pleasure undisturbed. The gold mines were the property of an ugly one-eyed race, who dwelt near a cave which is the home of the north wind, and when they found they were being quietly robbed, they consulted what they should do to punish the thief. It was not an easy task, for the griffin was much cleverer and quicker than his enemies, and, indeed, he nearly always got the best of it. Whenever they went out to dig for gold and emeralds, the griffin would hide until they had collected a large store, and then jump on them, flapping his great wings, and shaking his terrible claws, till they ran away in terror, dropping all their hard-earned treasure. There was only one way in which they could revenge themselves, and that was by carrying off the griffin's egg, that had the power of curing every disease from which mankind can suffer. But it was seldom that any one was fortunate enough or clever enough to win this prize, for the griffin is a very cunning creature, and more than a match for the one-eyed race. Still, now and then, an egg was discovered by some accident, and then how the whole nation rejoiced and prospered, till the precious thing got broken in some careless hands!

We all know about the battle, in 'Alice in Wonderland,' between the lion and the unicorn for the possession of the crown, and how the unicorn was worsted, and 'beaten all round the town,' by the victorious lion. Since that victory the lion has waved triumphantly from the English flag; but he and the unicorn are deadly foes still, and glare furiously at each other across the arms of

England. 'The unicorn and the lion being enemies by nature,' says a man who wrote three hundred and fifty years ago, 'as soon as the lion sees the unicorn, he betakes himself to a tree; whereupon the unicorn, in his fury, and with all the swiftness of his course, running at him, sticks his horn fast in the tree, and then the lion falls upon him and kills him.' The same story is told by other people, and this is what Shakespeare means when he says in one of his plays that unicorns may be betrayed with trees. There was only one way by which a unicorn could be taken alive, for 'the greatness of his mind is such that he chooseth rather to die,' one writer tells us; but this was a way which has been tried ever since the days of Samson, and even before him!

A beautiful young lady was dressed in her best clothes, covered with jewels, and seated in a lonely place in the middle of a forest to wait till the unicorn passed by; the hunters meanwhile lying hidden in a neighbouring thicket. By-and-bye a crackling would be heard among the branches, and after a little while the unicorn would come in sight, his sharp horn thrust out from his nose. Directly he saw the young lady he always went straight up to her, and laying his head on her lap, fell fast asleep. Then the hunters would steal out very softly, and throw ropes round the sleeping unicorn, and carry him off to the king's palace, sure of receiving much gold for their prize.

Living or dead the unicorn was held to be of great value for many reasons, but chiefly because his horn was used for drinking cups, and showed at once if any poison mingled with the wine. This was an excellent quality in times when people thought nothing of poisoning their nearest relations, and after the tiniest quarrel both parties went about in fear of their lives. The power of the unicorn's horn sometimes went even further, and dispelled the poison, for we read in an old chronicle of what happened in the waters of Marah, which Moses made sweet by striking them with his staff. 'Evil and unclean

HOW THE UNICORN WAS TRAPPED

beasts,' says the chronicler, 'poison it after the going down of the sun; but in the morning, after the powers of darkness have disappeared, the unicorn comes from the sea and dips its horn into the stream, and thereby dispels the poison, so that the other animals can drink of it during the day.' A few unicorns would be very useful on the banks of the rivers which water our manufacturing towns nowadays.

ABOUT
ANTS, AMPHISBÆNAS, AND BASILISKS

In the far-off country ruled by Prester John many wonders were to be seen, and among them hills of gold, 'kept by ants full diligently.' Now anybody who has studied the history of ants knows that there is no end to their ingenuity and cleverness; but they are not usually found as guardians of gold or precious stones. However, these ants were not at all like the little brown creatures we are accustomed to see, but as big as dogs, and very savage, thinking nothing of eating a man, and gobbling him up in one mouthful. So the people of the country found that if they wanted the gold they would have to obtain it by a trick, and began to watch and plan how to get the better of the careful ants.

Their chance came in the great heat of summer, as the ants used sometimes to fall asleep in the middle of the day. Then the people who had spies on the watch, day and night, collected hastily all the camels, dromedaries, horses and asses they could find, and loaded them with gold, and were off and out of danger before the ants, who were heavy sleepers, woke up. This did very well so long as the weather was hot, but when it grew cooler the ants worked hard all day, melting the gold in the fire; and then some other stratagem had to be thought of. One thing after another was proposed, but was rejected as being unpractical, till at last a man, who was cleverer than the rest, hit upon a way of turning a well-known quality of all ants against themselves. The industrious creatures could not bear to see anything standing empty or useless, and the treasure seekers, being aware of this, got together several mares, who had young foals, and placed on their backs empty vessels, which were open at the top, and reached nearly to the ground. As soon as the mares approached the hill, and began to graze, out

came the ants and began to fill the vessels. While this was going on, the foals had somehow been kept at a distance by the men, but as soon as they guessed the vessels to be almost full, they drove out the little creatures, who began to whinny after their mothers. At the sound of their cries, away galloped the mares, gold and all, and however often this trick was repeated, it never failed to be successful.

There is no time to tell of all the strange monsters that men used to invent just to frighten themselves with! There was a creature called the Odenthos, which had three horns instead of one, and felt a special hatred of elephants. There was the little Amphisbæna, which was something between a lizard and a snake, and had a head at each end of its body, so that it never needed to turn round. This must have made it very creepy to meet, but besides being horrid to look at it was very dangerous, as both of its heads were equally poisonous. Then there were yellow mice as large as ravens, and another kind as big as dogs, that must have looked rather like kangaroos, and a great many others, of which pictures may be found in old books. But none, not even the griffin or the unicorn, was as fierce as the small black basilisk, which was only a foot long. It got its name from a white mark on its forehead the shape of a crown, so they called it 'the king,' from the Greek word 'basileus.' It seems odd that such a tiny little animal could have caused such dread in men as well as beasts, but it really was a terrible little creature. It was enough for it to hiss for every living thing that heard it to scamper away to its den. If it spat, its venom was so deadly that rocks were rent by it, any bird that flew over it fell down dead into its jaws, and by merely looking at a man it killed the life within him. If he happened to come across a basilisk for the first time, and tried to cut off its head instead of running away, he fared no better, for the poison from its mouth would fly along the blade and cause his instant death.

We may wonder how, after a few years, there was anything but basilisks left on the earth, and perhaps there would not have been, but for the presence of weasels and of crystals. Weasels and basilisks had a natural hatred of each other, and rushed at each other's throats at every opportunity. The battle always ended in the same way, by the death of both combatants, for though 'the weasel overcomes the basilisk with its strong smell, yet it dies withal.' The piece of crystal was more useful still, for if you held it up between you and the basilisk and looked

THE DEMON OF CATHAY

through it the poison of the animal was driven back on itself, and killed the monster instead of the victim.

There are no basilisks nowadays, but their remembrance still lives in many of our proverbs.

The Demon of Cathay and his proceedings recall several of our old fairy tales, especially some of the Arabian Nights. He could talk the language of man and imitate any voice he chose, so that if he found a solitary traveller walking through a forest he would call to him by his name in the tones of some of his friends. The traveller would leave the path and go in the direction of the voice, when the Demon would spring out and devour him. Or he would mimic the roll of drums, or the blast of trumpets, and the poor man in surprise would think he must be drawing near a city, or at any rate approaching an army, so he would go in search of the sounds, only to find, when it was too late, that it was a trick of his deadly enemy's.

Quite as strange as the creatures on dry land were those that dwelt in the sea, for every animal that lived on earth had its fellow in the ocean. We read of sea-bears, sea-foxes, sea-asses, even of sea-peacocks; and now and then one might be found on the beach after a great storm.

Once some Dutch women, going down to the shore after a gale to see what they could pick up, were startled at finding a beautiful girl, with a fish's tail, lying among the shells and sea-weeds, beyond high-water mark. This was a mermaid, as anybody else would have known—a gentle creature, but without a soul. They took her home and taught her to spin and weave, and to kneel before a crucifix; but she was not happy, and always tried to escape into the sea. The Dutch women did not mean to be cruel, but they liked to have her there, and she was useful to them, so they kept a close watch upon her, and she lingered on in their house for fifteen years, fading gradually away, and dying in the year 1418.

On the opposite side of the North Sea, in the Firth of Forth, as well as in the Baltic and the Red Sea, sea-monks

FINDING A MERMAID

were at one time quite common, if we may believe a Scotch historian. Like their land brothers, they had a shaven spot on their heads, and wore robes and cowls; but instead of trying to help those who needed it, in one way or another, as land monks were supposed to do, they ate up everybody that came within reach. After this it is a comfort to think that a pair of shoes made from the skin of the sea-monk would last fifteen years!

Having once invented sea-monks, it was easy to go on and invent a sea-bishop, and pictures of him may still be seen in early books of travels with a crozier in his hand and a mitre on his head, and splendid vestments over his shoulders. He must have been a beautiful prize to catch, but he was very rare, and did not flourish out of the water. One was sent to the King of Poland as a present, but he pined away, and at length, finding himself in the presence of some bishops dressed like himself, he implored them by signs to release him from captivity. Overcome with pity for their brother in distress, they prevailed on the King to grant him his freedom, and when he heard the joyful news the sea-bishop at once made the sign of the Cross by way of thanks. The bishops escorted their brother solemnly to the sea-coast, and as he plunged beneath the waves he turned and raised two fingers, in the true form of episcopal blessing, and has never been seen on earth again, as far as we know!

DRAGONS

NEARLY a thousand years ago there lived a historian who set down in his book not only accounts of real battles and sieges, but also a strange medley of other facts besides. Of course he thought all he wrote was true, for history, as the dictionary tells us, is 'an account of facts and events,' and the business of the historian is to write about them. The stories in this old book about magic, spells, dragons, and monsters may, perhaps, make us smile nowadays, when we are taught that fairy rings are not caused, as we should like to suppose, by the good people, but by 'an agaric or fungus below the surface which has seeded in a circular range.' But it must be remembered that to the men of old time all these matters were very real. Our historian, in common with many wise men who lived hundreds of years after him, believed without doubt that the world was full of strange creatures which lived in pathless woods, in rivers, on mountains, or in the sea. One of his tales is the description of a voyage by King Gorm Haraldson, under the guidance of Thorkill the Icelander, in quest of treasure supposed to be guarded by Giant Garfred, who lived in a 'land where no light was, and where darkness reigned eternally.' 'The whole way was beset with perils, and hardly passable by mortal man;' nevertheless, three hundred men declared their willingness to follow the King and make the attempt. After many adventures the wind took them to Utter Permland, a region of eternal cold and deep snows, full of pathless forests, haunted by dreadful beasts. King Gorm and his followers were met by a huge man named Gudmund, the brother of Giant Garfred, who gave himself out to be the guardian—the most faithful guardian—of all men who landed in that spot. In reality he was a treacherous scoundrel, but at the outset he invited them to be his guests, and 'took them up in carriages.' 'As they went forward they saw a river which could be crossed by a bridge of gold. They wished to go over it, but Gudmund restrained them, telling them that, by this channel, Nature had divided the world of men from the world of monsters, and that no mortal track might go further.' Well, here we take leave of King Gorm and Gudmund, and we will cross in imagination that golden bridge into monster-land, though they did not, nor does our historian, give any particular description of the monsters which lived there; but, from other ancient writers, we can get a pretty fair idea of what he would have been likely to say about them if it had suited his purpose. He would certainly have included a stray dragon or two; indeed, elsewhere, he does actually give us two dragon-slaying stories, the first of which concerns King Fridleif, who was wrecked on an unknown island.

He fell asleep, and dreamt that a man appeared before him, and ordered him to dig up a buried treasure, and to attack the dragon that guarded it. To withstand the poison of the creature, he was told to cover himself and his shield with an ox-hide. When he awoke he saw the dragon coming out of the sea, but its scales were so hard that the spears thrown by Fridleif had no effect, and the only thing that happened was the uprooting of several trees by the monster, which wound its tail round them in a fit of temper. However, the King observed that by constantly going down to the sea the dragon had worn a path, hollowing the ground down to the solid rock to such an extent that a bank rose sheer on each hand; so

Fridleif seems to have lain in ambush, as it were, in this hollow channel, and to have attacked the creature from beneath, where its armour was less proof against assault; in this way he slew it, unearthed the money, and had it taken off in his ships.

The second story concerns another King, called Ragnar Lodbrog, which means Ragnar 'Shaggy-Breeches.' This is how he came to be known by his nickname, which was bestowed upon him by Herodd, King of the Swedes: Ragnar was in love with Thora, Herodd's daughter, who had received from her father two snakes to rear as pets. They had given to them daily a whole ox upon which to gorge themselves, so they ate and ate, and grew and grew, until at length they became a public nuisance, so huge were they, and so venomous withal that they poisoned the whole country-side with their breath. The Swedish King repented his unlucky gift, and proclaimed that whosoever should remove the pests should marry his daughter. Many tried and perished; but Ragnar was now to prove himself the hero. He asked his nurse for a woollen mantle, and for some thigh-guards that were very hairy; he also put on a dress stuffed with hair, not too cumbersome, but one in which he could easily move about. He took a sword and spear, and, thus accoutred, fared forth to Sweden. When he arrived, he plunged into some water, clothes and all, and allowed the frost to fashion for him, as it were, a coat of mail, impervious to the venom of the snakes. Leaving his companions, he went on to the palace alone; then the combat began. An enormous snake met him, and another, as big, crawled up to help its companion; they belaboured Ragnar with their tails, and spat venom at him from poisonous jaws. Meantime, the King and his courtiers 'betook themselves to safer hiding, watching the struggle from afar, like affrighted little girls.' Ragnar, however, persevered, his frozen dress protecting him from the poison, and with his shield he repelled the attacks of the snakes' teeth; at last, though hard pressed, he thrust his spear through the creatures' hearts, and his battle ended in victory. A great banquet was held in the palace; Ragnar received at once his bride and his nickname of 'Shaggy-Breeches,' as we have seen. He did many other brave deeds, and was a successful rover; but was cruelly put to death by an English King called Ella, who threw him into a pit full of snakes. Ragnar's device of freezing himself into a suit of ice

RAGNAR DOES BATTLE WITH THE SERPENTS

armour recalls to us a similar plan adopted by a race of monsters universally believed to have lived in Africa; nearly all the old writers of marvels allude to them, under the name of 'Cynocephali,' which means 'dog-headed,' that is to say, their bodies were those of men and women, but their heads were the heads of dogs. They lived upon goat's milk; but although that seems to mean that they dwelt quietly amongst flocks and herds, they seem nevertheless to have been fond of a fight whenever there was the least chance of war with neighbouring tribes. To prepare for battle, like Ragnar, they jumped into water, and then rolled themselves in the dust until their bodies were covered with it; then they allowed the sun, which, of course, is always very powerful in Africa, to bake it into a sort of cake or mud-pie crust, which formed the first layer of defensive armour; when that was sufficiently dry and hard they repeated the process, not once or twice only, but again and again, until they thought their coat of mail, if we may so call it, strong enough to be proof against the arrows of the enemy.

A very worthy writer, who lived about 1600, has told us that he quite believes in the reality of winged dragons. After giving us some wonderful stories about them, he remarks that 'from these and similar tales we can easily see that what we find in *other* authors about winged dragons is all true.'

Switzerland, especially that part of it round about the Lake of Lucerne, was famous for these creatures. There is opposite to the town of Lucerne a mountain, called Pilatus, from the tradition that Pontius Pilate, when banished by the Roman Emperor Tiberius, wandered there, and threw himself into a black lake at the summit. His ghost is supposed to haunt the place; once a year it appears, clothed in robes of office, and whoever is unlucky enough to see it, will die before the year is out. Mount Pilatus often has on a cap of clouds, and it is said that the weather will be fine, or the reverse, according as Pilatus has his cap off or on. We may well imagine it, therefore, to be a wild, eerie sort of place, in every way suitable for dragons to take up their abode. Our old author then tells us that a peasant one morning was mowing hay; he looked up, and at that moment there issued from Pilatus a huge dragon, which flew across the lake to a mountain on the other side. In its flight there dropped from it something which the peasant could not clearly distinguish, for he was too frightened to observe accurately, and indeed was nearly fainting; but when he recovered, he found in a meadow a mass of what appeared to be solid blood. Enclosed in this was a stone of many colours; this stone turned out to be of priceless value, for it was a certain cure for every disease under the sun ; and more especially for such as were caused by poison or bad air of any kind ; it was still in Lucerne at the time the author wrote.

Another man of that city, called Victor, saw a still stranger thing on Mount Pilatus. He was a cooper by trade, and one day, when out looking for wood wherewith to make his casks, he lost his way in the recesses of these Alpine rocks and forests. All day long he wandered about, until, at twilight, as he was just about to lie down and rest, he fell into a deep chasm which, owing to the failing light, he had not noticed. Fortunately he fell into some soft mud at the bottom, but though he broke no bones, he fainted. When he recovered, and began to look round, he discovered that there were absolutely no means of escape. The hole was as deep as a well, with steep sides which could not be scaled. Stretching along the whole length of this cavern, and on either side, were other tunnel-like openings, a succession of smaller caves; into one of which he was about to enter when, lo! two dragons came forth from it, and he supposed that his last hour was at hand. The creatures, however, offered him no violence ; they were inquisitive, it is true, wondering, no doubt, what sort of new companion this was, who had found his way into their dwelling ; but all they did was to rub themselves against the man's body, caressing him, as it were, with their long necks and with their tails, just like a purring cat. For six months Victor lived in this underground cavern. 'But what did he live on?' you may ask, with Alice, when the Dormouse told his story of Elsie, Lacie, and Tillie in the well. These three sisters, you may remember, lived upon treacle, which was sweet, if unwholesome ; but the Lucerne man's diet was even less satisfying, being only the moisture which trickled from the surface of the rock. Learned men have certainly proved that it is possible to keep oneself alive for many weeks without food, if a sufficient supply of water be taken; but I do not remember to have met with any other case where any one lived for six months upon such provender. When spring came round the dragons thought it time to leave their abode; unfolding its wings, the first one flew up, and the second was preparing to follow, when Victor, seizing at once his opportunity and the tail of the dragon, was carried by the creature into the upper world. He found his way back to Lucerne; but a return to his ordinary food, of which he had been for so long deprived, brought on an illness, and in two months he died. His adventures were embroidered upon an ecclesiastical vestment, which used to be shown in the church of St. Leodegarus to any sight-seers who might wish to see it.

Near the church of St. Stephen in the city of Rhodes there was a vast rock, and a cavern in it from which issued a stream of water.[1] In this subterranean cave there lived, in the year 1345, a terrible dragon, which devastated the whole island; not only did it devour sheep, cattle, men, anything living, upon which it could seize, but its breathing was so pestilential that the very atmosphere was poisoned by it. Nobody could venture to go near the part of the coast where it dwelt; in fact the Grand Master of the Knights strictly forbade anybody belonging to the Order to attempt it, under this severe penalty: First, he was to suffer the disgrace of

[1] The Knights of St. John of Jerusalem, or the Knights Hospitallers, as they are sometimes called, were an Order founded in the eleventh century, some time after the first crusade: in the fourteenth century they took the Island of Rhodes, in the Mediterranean, and held it against the Turks. It was during their life in this island that the events occurred which are now to be described. The account is taken from a history of the Order, which is quoted word for word by the author who has told us the story of the Lucerne dragons.

being deprived of the marks and dress of the Order; and, secondly, his very life was to be forfeited. Nevertheless there was a young Gascon Knight, of noble birth and great courage, who was not to be deterred from his project by this edict; on the contrary, he thought an opportunity presented itself of winning much honour and renown. His name was Deodatus de Gozon. He kept his own counsel, telling nobody in the city of his plan, but he went to the Grand Master and begged leave of absence on the pretext of business at home. Having got leave he went into the country to carry out his design; but he was careful, before starting, to observe the dragon as closely as possible, so as to remember every point in its horrid carcass. What he saw is thus described: It had a body as thick as that of a carthorse; its long and prickly neck ended in a serpent's head, which was provided with long ears like those of a mule; its mouth gaped widely open, and was furnished with the sharpest of teeth; its enormous eyes shone so brightly that they seemed to emit flames of fire; and its feet (of which it had four) were armed, like bears' feet, with sharp claws. In its tail and other parts of its body it resembled a crocodile, wearing an armour of the hardest scales cunningly disposed; from its sides issued two gristly wings, in colour not unlike a dolphin's gills—the upper surface blue, the lower a sort of reddish yellow, this last being the general hue of its entire body. Swifter than a horse, when it moved abroad in search of food, it did so partly by flying, partly by running; its scales, too, made such a clattering, as of crockery, and its hissing was so terrifying that people at a great distance were almost frightened to death.

De Gozon, accordingly, having looked carefully at the monster, as we said, withdrew into the country, where he set to work and contrived a creature exactly like the dragon in every respect; he made it of paper and stuffed it with tow; then he bought a well-trained charger, and a couple of English dogs—bull-dogs, in all probability. He now taught his servants how to make the tow dragon imitate the movements of the real dragon; that is to say, they snapped its jaws, and made it lash its tail about and flap its wings; all this they did by means of ropes. Next he mounted his horse and brought his dogs into action, setting them at the sham dragon, and exciting them with cries, until their rage knew no bounds; hardly did they set eyes upon it, when they flew at it to tear it in pieces. These exercises went on for the space of two months, at the end of which De Gozon, considering his men and dogs sufficiently well drilled, returned to the city. Arrived there he lost no time in carrying out his project; arming himself with breastplate, lance and sword, he went to the church of St. Stephen, which was near the monster's den, and prayed, devout knight as he was, that his enterprise might be crowned with success. He then gave particular instructions to his servants as to what they were to do: they were to watch the battle from a lofty rock, and if the creature won, they were to escape as best they could; but if he slew the dragon, they were to hasten to his aid, for it was only too likely that even victory would cost him dear, and that he would stand sadly in need of such remedies as they could bring.

All was now ready; so the Knight, entering the cave, began to screech and yell lustily in order to wake up the dragon and annoy it; then, rushing out himself, he mounted his charger, and awaited the attack on a piece of level ground. He did not have long to wait; scarcely was he mounted when the sound of the well-known hissing was heard, and the clattering of the huge plate-like scales warned him that the monster was after him in full cry—and, indeed, as it came at him, partly running, partly flying, the creature itself thought it saw in the bold Knight an opportunity not lightly to be missed; for all was grist that came to its mill—flocks, herds, horses, and men, as we have already seen. De Gozon hurled his spear at the beast, but the shaft shivered into a hundred pieces against the hard scales, so that, thus early in the fight, he lost the use of one of his best weapons. But the dogs now made a diversion in his favour, for by worrying the monster on this side and on that, they so engaged its attention that the Knight had time to dis-

DE GOZON AND HIS DOGS FIGHT THE DRAGON

mount, and make ready with sword and shield for a combat on foot. Rearing itself up on its hind legs, the dragon endeavoured, as a bear will do, to hug its enemy to death, but it now exposed the under surface of its neck (which was comparatively unprotected by scales) to the attack of De Gozon. In an instant he thrust his sword into its throat; a deluge of blood gushed out; the monster tottered, and fell; but in its fall crushed to the ground the brave Knight, who was already sufficiently wearied with the strife, and half poisoned besides by the dragon's noisome breath. The servants, however, seeing the dragon fall, rushed down from the neighbouring heights, and thinking they could discern some faint signs of life in their master, filled their caps with water from the stream hard by, and dashed it over him. He soon recovered sufficiently to be able to mount his horse and ride back to the city, where he told the Grand Master of his splendid exploit, thinking, not unnaturally, that honour, reward and glory would be his—who had freed the country from such a dire pest. But, alas! the Grand Master set the duty of obedience before even such deeds as De Gozon's. The Knight had disobeyed the edict, had been altogether far too foolhardy and presumptuous, and must take the consequences; he was accordingly degraded and imprisoned. Not for very long, however, we are happy to think, for the tidings soon spread over the whole island, and people were so strong in his favour, that the Grand Master was induced to relent. De Gozon was liberated from prison and reinstated. Shortly afterwards all the people in the city assembled to do him honour in a procession; nor were the brave dogs forgotten, for had it not been for their furious onslaught it is not likely that the Knight would have lived to tell the tale. They were led at the head of the procession, with the dragon's skin borne before them, heralds proclaiming as they went: 'These are the brave English dogs, the preservers of the Knight, the conquerors of the dragon.' Four years afterwards the Grand Master, Elio de Villanova, died; and Deodatus de Gozon was unanimously elected as his successor--in the year 1349.

THE STORY OF BEOWULF, GRENDEL, AND GRENDEL'S MOTHER

LONG, long ago, perhaps nearly a thousand years before the adventures of the Knight of Rhodes of whom you have just heard, there lived a King of Denmark called Hrothgar. That is a curious name, you may think; but you can recognise it in our own word 'Roger,' which, of course, is common enough. This King lived in a palace, called Heorot, a princely abode, beyond what the sons of men had ever heard of; he had a beautiful wife called Waltheow, and gold, silver, and riches in abundance were his; moreover as his knights, earls, and retainers were all devotedly fond of him, he seemed to have everything in the world which could make him happy. In those days, when feasts were being held in the great halls, it was customary for one who was called a 'skald'—that is, a poet or minstrel—to sing or recite poems before the assembled company. On one of these occasions the 'skald' made poems about all sorts of evil things, wicked spirits, demons who abode in darkness, giants, ghosts, and sin and wickedness generally. It was, perhaps, not quite the sort of song to make merry the hearts of the feasters, and, in fact, it had the opposite effect, for they broke up ill at ease, as if some deadly peril were in store; nor were their presentiments without reason. That night there came to the Palace a monstrous and superhuman being named Grendel, who was the very incarnation of all cruelty and malice. He was a creature of enormous strength and size ; for we read later in the story that it required four men to carry his head when he was dead. He lived an evil life, and wandered about, a lone dweller in moors, marshes, and in the wilderness. Savage and fierce as he was, nothing exasperated him more than that the King and his people should be so happy ; the sound of joy and revelry within the Palace was to him as gall and wormwood. That very night, therefore, when the skald recited his ominous poem, Grendel left his fens and marshes, and came silently to the Palace, where he found the Danes all asleep. Thirty of them he killed, devouring fifteen in the hall itself, and carrying off the rest to the marshes. Despair there was and lamentation in the morning when the other Danes arose from sleep ; but none knew, or could even suggest, what was best to be done. For twelve years were the people grievously afflicted by the cruel Grendel, ' the grim stranger, the mighty haunter of the marshes, the dwelling of this monster race.' He persecuted them right sorely, nor would he have peace with any man of the Danish power. A dark, deadly shadow, he attacked alike tried warriors and youths, he ambushed and plotted, roaming the night long over the misty moors, contriving evil in his heart continually.

Matters, then, were at this pass, when a neighbouring King called Hygelac heard of the Danes' misfortunes. Hygelac reigned over the Jutes in Gotland, and he had a nephew called Beowulf, who, in common with the King and the rest of the people, was distressed to think of Hrothgar's troubles. So Beowulf made him ready a good sea-boat, took fourteen of the bravest men-at-arms as his comrades, and set sail to help Hrothgar and the Danes. When the Danish King was told of Beowulf's arrival, he was, as you may well suppose, only too delighted, and hailed him as a heaven-sent champion, for he already knew all about him, how valiant he was, and how strong ; ' for,' said Hrothgar to his people, ' it used to

be said by seafaring men that this fearless warrior had in his grip the strength of thirty men.' When Beowulf came before Hrothgar, he told him, what the King already knew, that often before he had encountered sea-monsters, destroyed the Jotun tribe and slain night Nixes; and that hitherto all his deeds of prowess had been successful. 'I hear,' he said, 'that Grendel, from the thickness of his hide, cares not for weapons; I therefore disdain to carry sword or shield into the combat, but with hand-grips will I lay hold on the foe, and fight for life, man to man.' Beowulf ended by asking that his 'garments of battle' might be sent back to his lord and kinsman Hygelac, if Grendel proved victorious in the fight. The King relied with steadfast faith upon his guest; there was now joy in the Palace of Heorot, and Queen Waltheow herself, golden-wreathed, came forth to greet the men in the hall; to each she gave a costly cup—to each his several share—'until it befell that she, the necklaced Queen, gentle in manners and mind, bare the mead-cup to Beowulf,' and thanked God that she might find any to trust to for relief in her troubles. They all retired to rest; but not one of Beowulf's comrades thought that they would escape alive, or get them thence in safety to their well-loved homes.

That night from the moor, under the misty slopes, came Grendel prowling; in the gloom he came to the Palace, where the men-at-arms slept, whose duty it was to guard the battlemented hall; they slept, all save one. With his vast strength the monster burst open the door, and strode forward, his eyes blazing like fire. With a grim smile of delight he saw the sleepers, seized one of them and devoured him all but the feet and hands. Then he reached out at Beowulf, but the warrior clasped the extended hand and firmly grappled with the enemy. A battle royal ensued; the hall resounded with cries and shrieks, for the Danes were roused from their slumbers. They tried to help Beowulf with swords and other weapons, not knowing that they were of no avail against the monster. But the Jute yielded never a whit, he pressed Grendel harder and harder with that mighty hand-grip of his, and by sheer strength tore off the monster's hand, arm, and shoulder. Grendel fled; back to the lake he went, to the Nixes' mere, where the water for days afterwards was troubled and discoloured with blood.

As for Beowulf, the grateful King could hardly thank him enough. A feast was prepared, the walls of the great hall were covered with cloth of gold, and the hero received a war-banner, helmet, and breastplate, besides golden cups, a superb golden collar, and many other precious things. When the banquet was over they all retired to rest, as they supposed, in safety. But an avenger was at hand, Grendel's mother, a monstrous witch, ravenous, wrathful, and cruel as her son. She burst into Heorot, seized the man who was the King's favourite amongst all his nobles, and carried him off to the lake. She also took with her Grendel's blood-stained hand, which had been put up as a trophy. Beowulf was not in the Palace at the time, for another lodging had been given to him; but he was quickly summoned after this new disaster. 'Never fear,' said he, 'I promise thee she shall not escape, neither by water, nor into the earth, nor into the mountain forest, nor into the bottom of the sea, let her go where she will.' So they made ready at once to go to the lake, which was about a mile from the Palace; a gloomy water it was, overhung with trees, and how deep none had ever found out; every night, men said, a strange fire was to be seen on its surface, so none cared about going there. However, the King's horse was now saddled, and his men-at-arms were ready; Beowulf put on armour to protect his body from the enemy's grip, and a white helmet guarded his head. One of Hrothgar's men lent him a short sword that had never

GRENDEL'S MOTHER DRAGS BEOWULF TO THE BOTTOM OF THE LAKE

yet failed anyone who had used it in battle. Then the expedition started: over a steep and stony rise through narrow roads, past precipitous headlands they went, till they came to a bare rock and a cheerless wood, below which lay the water, dreary and troubled. They were maddened with rage when they saw the head of Æschere lying on the ground; he was the noble taken by Grendel's mother.

The water of the lake was bubbling with blood; many strange creatures of the serpent kind glided over the surface, and the men could also see Nixes lying on the headland slopes. Beowulf shot at one of the horrid water creatures with an arrow, wounding it only; but the King's men pursued it with poles and battle-axes, and killed it. Then Beowulf asked Hrothgar to send back all his presents to Hygelac, if it should happen that he, Beowulf, perished in the water. Hastening away, he plunged into the lake, and it was not very long before Grendel's mother found out that some man from above had invaded her dwelling. She grappled with him in her dreadful grasp, endeavouring to crush him to death, but his chain-mail protected him. Then she dragged him down to her den at the bottom; but meanwhile many strange beasts with terrible tusks pressed him hard in those depths, one of them even rent his war-shirt with its talons. Beowulf found himself in some kind of dreadful hall, where no water seemed to touch him; the light of a fire, a glittering ray, lit up the cavern. He could now clearly distinguish the mighty lake-witch, and thrust strongly at her with his war sword, which rang out shrilly on her head. But, alas! its edge would not bite; she had probably bewitched it with spells, as often happened in old days. So Beowulf threw away his sword, and came to close grips with her, trusting in his mighty strength. He seized her by the shoulder, but unluckily tripped and fell. In a moment she was upon him, and seized her broad dagger with deadly intent. Then, indeed, had it gone hard with Beowulf but for his coat of chain-mail, which protected his shoulder from the furious blow she gave. Suddenly he saw lying on the floor a magic sword; a huge weapon with finest edge, forged of old in the time of the Jotuns, or giants, whose work it was. No ordinary man could have wielded that blade, but Beowulf seized it, and smote the witch a fearful blow, almost cleaving her body in twain. A bright light shone up at once in the cavern, which the warrior now began to explore; nor had he gone far before he found Grendel lying on a couch, dead, so Beowulf cut off his head. Meanwhile Hrothgar and the rest of the Danes had been sitting watching the water, which suddenly became thick and stained with blood; they had no hope that Beowulf survived. What, then, was their astonishment and delight to see him swimming towards them, breasting the waves with mighty strokes, and bearing the head of Grendel with him. And now a marvel befell; the sword with which Grendel's mother had been slain began slowly to melt away, just like ice; for the hag's blood was of such power that it consumed the blade, until nothing was left but the hilt, which was of gold, richly chased, and carved with strange characters called 'runes.' Beowulf swam ashore, and gave an account of his adventures; four men, as we have already said, bore Grendel's head to the Palace, where the hilt of the magic sword was closely examined. The characters graven upon it were found to be a description of the battle between the Gods and the Frost-Giants, in which the Giants were defeated and overwhelmed in a flood. There is an account of it in an Icelandic poem, called the 'Völuspa,' or the 'Song of the Prophetess,' which describes the Northern ideas of the creation of the world; and tells how evil and death came upon man, predicts the destruction of the universe, and gives an account of the future abodes of bliss and misery. Thus did Beowulf deliver the Danes from their misfortunes, after which he returned home, and on the death of his uncle, Hygelac, became King of Gotland.

THE STORY OF BEOWULF AND THE FIRE DRAKE

BEOWULF was a wise King, and had ruled his country well for fifty years, during which nothing had happened to mar the happiness of him or his subjects. But now trouble was about to arise. Hidden away in a mound of earth was a vast store of treasure, gold, silver, jewels of great price, and this hoard for three hundred years had been guarded by a monstrous Fire Drake. One night, while this dragon slept, a man succeeded in entering the storehouse, from which he stole a cup and many valuable jewels. When the serpent awoke its rage knew no bounds; it came forth from its cave, endeavouring to track the man, whose footsteps it could see on the shore, but without success. So it waited till evening, vowing that many should pay dearly for that drinking-cup. Then again it came forth, wandered all over the country at night, setting every house it could see on fire, for its scorching breath and the brands it carried with it were irresistible. Beowulf's own home, in common with others, was destroyed, whereupon he bethought him of vengeance, remembering how of old he had been successful in quite as dangerous undertakings, and how he had outlived every quarrel, every perilous enterprise. Knowing well that no ordinary defence would avail him anything against the Fire Drake, he had fashioned for himself a curious battle shield, all of iron. Choosing eleven companions, he went to look for the dragon; the way was hard to find, so the man who had been the cause of all the mischief went with the little band as a guide: indeed, he was the only one who knew where the dragon's hoard was to be found; besides, he was very much ashamed of himself, and was anxious to do all in his power to atone for the disasters which his theft had brought about.

When they arrived at the Fire Drake's lair, which was near the sea, they saw an arch of stone, and a stream issuing out of it from the mound. The water was so hot, by reason of the dragon's flame continually beating upon it, that a man could not bear his hand in it for any length of time. Beowulf told his companions to wait outside, whilst he himself went into the cave. The Fire Drake, hearing his footfall and his voice, knew at once that an enemy was near, so it coiled itself up ready to spring to the attack. Blazing like a live coal, it advanced with a rush, Beowulf defending himself as best he could with his shield. He dealt the monster a terrible blow with his sword, which, however, failed to hurt it, indeed, it only roused it to greater fury. Breathing flames the Fire Drake pressed the valiant King to the utmost extremity, and it seemed as if it was to go ill with him that day. His companions, too cowardly to help him, watched the combat in terror, crouching down in the wood near by to save their lives. Yet there was one among them, Wiglaf by name, who plucked up courage to try to help the King, for he remembered how kind Beowulf had been to him in former days, in granting him a wealthy manor, and other favours, and besides, he was in a way related to him. So this brave young warrior grasped his shield of yellow linden wood, and drew his sword, rushing through the smoke to help his liege lord. 'Dear Beowulf,' cried he, 'have courage; remember how thou did'st say aforetime that glory should never depart from thee; now must thou defend thy life to the uttermost—see, I come to help thee.' On rushed the serpent against its new

adversary; from its body and mouth issued many coloured flames, which burnt up Wiglaf's wooden shield, so that for protection he crouched under the iron shield of Beowulf. The King now struck with all his force at the dragon, but, alas! his good old sword shivered in pieces; and now for the third time the monster rushed at him, and succeeded in encircling his neck in its horrid coils. Still, the King's hands were free, so that he could draw a dagger which he bore on his corselet; Wiglaf, meanwhile, was also hewing at the creature, and before long Beowulf was able to stab it to death. Thus they slew the Fire Drake; but Beowulf had received a deadly wound, which soon began to burn and swell, and though Wiglaf brought him water and tended him with all affection, the King felt his end to be near. Anxious to know of what the treasure consisted, he sent Wiglaf into the cave to explore it. Riches of all descriptions were discovered— jewels, gold, handsome bowls, helmets, armlets, and, most curious of all, a gilded standard, which was flapping over the hoard. From this standard there came a ray of bright light, by which Wiglaf could easily see around him. Nothing was to be seen of the dead Fire Drake, so Beowulf's messenger plundered the hoard at will. He piled up bowls and dishes in his bosom, took the standard, and a sword shod with brass, hastening with them back to the King, who, he was half afraid, might die during his absence. Beowulf was alive, however, though in sorry plight, so Wiglaf fetched more water wherewith to refresh him. Then spake the brave old King his last words on earth, the while he looked sadly on the gold: 'I give thanks for these beautiful things, which here I gaze on, to the Lord of all, to the King of Glory, the eternal Lord, for that I have been able before my death-day to gain so much for my people. Fulfil ye now with this hoard my people's needs, for here I may no longer be. Let the warriors build a mound at the headland which juts out into the sea. Rear it that it may tower high up on Hronesness, and so perchance my people may bear me in mind. Yea, let it be for a landmark to seafaring men, who may call it Beowulf's Mound— a beacon of safety for such as are in stress on the storm-tossed sea.' Thus died Beowulf. When the news spread the people flocked out in hundreds to the spot where the fight took place. Sadly they looked on the lifeless body of their chief lying on the sand, and with astonishment they saw the carcass of the Fire Drake, full fifty feet long, and the hoard of treasure beside it. They loaded the treasure on a wain and bore it away; the dragon's body was pushed over the cliff into the sea. Then they made ready a vast funeral-pyre for their beloved King, even as he had wished. Black over the blaze rose the wood smoke; while sad and dejected in spirit sat the people, mourning their lord's fall, bewailing the death of him who among world Kings had been the mildest, the kindest of men, and the most gracious to his people.

A FOX TALE

WHEN I ask children to tell me what they know about a fox, they almost always reply: 'He is a little red beast, very cowardly and cunning: he kills hens, and has a very bushy tail.'

This is all quite true; but Renard lives a very hard and extremely uncertain life; yet all the while is so dashing and gentlemanly, so quick and clever, that you must forgive him one or two faults.

He begins his life in a nice warm nest of hay, dry leaves and moss, at the bottom of a deep burrow, generally in a sandy bank. His mother tends him, fondles him, plays with him, as only a mother can; her one ambition being to keep him concealed from human sight. Once a man came by a particular burrow with his dog, hung about for some time near by, and then went away again. That night, Mother Fox took her little one up in her mouth by the nape of his neck, and set off to find a safer home. Hardly had she gone ten yards from her burrow when a dog jumped out of some bushes and gave chase.

Mother Fox flew like the wind over hill and dale, on and on, till her breath began to come in short, sharp gasps, and she felt she would soon have to turn and face her pursuer. But never once did she dream of dropping her little one and thereby saving herself; oh, no! cowardly as foxes are ever said to be, the mothers will alway die fighting for their young.

Happily for this mother, however, a long stretch of

whin bushes just then hove in sight, and, summoning up all her strength, she made a last spurt, and crept into the thick of them. The dog followed for a short distance, but evidently found the thorns too sharp for his thick nose and long flapping ears, for he soon retired, leaving Mother Fox gasping, but triumphant, with her little one safe and sound. She crept some way farther into the bushes to guard against pursuit, and there lay hidden till nightfall, when once more she stole stealthily out with her cub in her mouth, and made tracks for a hollow tree which she knew of in the neighbourhood. Reaching it in safety, she soon had a warm nest made in the dark recesses of the tree trunk, where little Renard lay for weeks eating and sleeping by turns, till he grew into quite a respectable fox. And what a merry little fellow he was! As playful as a kitten, and quite as active; gambolling all round and over his poor patient mother, burying his face in the furry depths of her brush, or, if she refused him that huge enjoyment, flying round and round in a mad race after his own, till he looked for all the world like a woolly spinning top!

But life is not all play, even to little foxes, and young Renard was awakened every night by a poke in the back from his father, who wanted his company on all nocturnal expeditions; for, strange as it may seem to us, foxes have lessons at night and sleep through the day, instead of having lessons through the day and sleeping at night. And sometimes little Renard was good at his lessons, and sometimes he was not. Very often, on catching sight of a pheasant or a partridge, instead of trailing his hind legs out behind him, as his father did, he would forget, and gallop full tilt at his prey, and yelp with excitement, expecting the bird to sit still and be caught! and not till the pheasant was whirring away high in the air would he remember that stealth and cunning alone will win a fox his daily bread.

Hitherto little Renard had known no sorrow, and it

came to him very suddenly one night when he was out foraging with his father. They were creeping along together, keeping as much under cover of the long grass as possible, when Mr. Fox struck on a hare's trail, and off the two set with their noiseless gliding motion, their noses well to the ground, and their ears alive to every sound under the moon. All at once, when Mr. Fox was slinking under a gate, he began to back and wriggle as if trying to escape from some unseen power. Young Renard pulled up, watched the old fox anxiously for a moment, and then, seeing a dark form approach, he fled, thinking only of the safety of his own red skin.

Truth to tell, it was a poacher's net into which the old fox had fallen, and the more he struggled to free himself the tighter he became entangled. Instinctively feeling this, and hearing the poacher himself approaching, the cunning creature lay perfectly still in the hope, no doubt, of escape by feigning death. But the wary old netter was quite up to Renard's tricks; and seeing that his nets would be torn to pieces if he did not free the animal at once, he tried to loosen one end off the gate. Mr. Fox, however, thought the trap had been set for *him*, and was determined not to be taken in that way; so he snarled and bit at the man every time he came near the gate. Again and again the poacher tried, but at last, losing patience, he seized some heavy stones off a dyke close by, and pelted Mr. Fox till he died. 'And,' said the poacher afterwards, when telling the tale to his friend, 'it went sore against me killing that animal, for never a sound did it make from first to last.'

Young Renard had witnessed his father's fate from a safe distance, and ran off as soon as all was over to tell his mother. He found her busily scratching up their morning meal from the various larders round about: for foxes, you know, always bury their prey, and never keep more than one 'joint' (be it of bird or beast) in the same larder at the same time; they have game safes scattered for miles round in all directions, so that if one is discovered, they still have two or three other breakfasts or dinners waiting for them somewhere else.

Mrs. Fox did not seem to take her loss very much to heart— merely told young Renard that he would have to cater for himself and her now, and bade him hurry on with his breakfast.

His meal over, Renard sauntered about till he found a cosy place in a spruce covert wherein to rest. He tried this place and that, but none suited him: one was too humpy, another too deep, and a third full of pine needles; but at last, after a great deal of thinking and poking, he twisted himself into a round woolly ball, curled his tail over his nose and slept soundly till dusk.

When he awoke, he remembered with a pang that he would have to do the hunting all alone that night, and for every night to come; and that, if there were any poachers' nets or gamekeepers' traps, he would be sure to fall into them, as now he had no one to reconnoitre on ahead.

He thought over all the birds and beasts which he liked best to eat, and decided that a nice fat chicken was really dearest to his heart. So away he went, as soon as it was dark, to a farmyard some five miles off. Arrived there, he was not long in discovering the hen-house, and, luckily for him, the hen-wife had left the small lower door open to admit three stray ducks who had not made their appearance at the usual locking-up hour. Renard was not slow to avail himself of this piece of good luck, and, creeping slyly through the hole, stood quite still for a minute or two to see if his entrance had been observed. It had evidently not, for there was the silence of sleep upon the unsuspecting fowls; so, cautiously, and with a beating heart, he softly scaled the ladder, and crept towards an open coop which was standing on the floor. There was a nice fat chicken inside, which stirred a little as Renard approached, and fearing it was going to wake up and cackle, he made a dash and grabbed it by the neck. The chicken struggled fiercely, one of its wings got caught in the bars of the coop, and the scuffling that ensued soon woke the whole roost. Then began such a cackling, and screaming, and quacking as Renard had never heard before, and he tugged at his chicken in a perfect frenzy of despair, expecting the hen-wife to appear every minute. At last he got free of the coop, and was just going to descend the ladder when the door opened, and a woman came in with a lantern. Renard saw in a moment that escape by the door was impossible, and instantly his fertile brain had planned a bold scheme. Still holding the chicken in his mouth, he stumbled on the top step of the ladder and rolled heavily to the bottom. The hen-wife ran forward, stick in hand, to put an end to the thief; but seeing he lay quiet in a huddled-up heap, she seized his tail, and dragged him towards the door. Imagine the shock poor Renard experienced when he felt his beautiful brush grasped by the sturdy hen-wife's fingers! and the terrible longing which came over him to turn and rend his captor. He restrained himself, however, when he saw he was being dragged towards the door; and when the hen-wife, feeling his stiff and lifeless body somewhat heavy, tumbled him into a thicket of nettles, he almost barked with delight. True, he had lost his chicken, but had gained in cunning, and cunning is honour among foxes.

Renard's exploits are too many and various to mention; but there is just one more you must hear about, because it shows he had pluck, as I think all foxes really have.

He was slinking along at dusk through some long grass, close in to a wood, when, snap! bang! and Renard was fast in a trap, caught by the leg. He tried dragging, pulling, and shaking it all in vain; the trap clung to his flesh with its iron teeth, and would not let go. After persevering for an hour or two, Renard gave up those methods, and tried another, beginning deliberately to gnaw off his own leg! Who shall say now that foxes have no courage? In a few minutes he was free of the trap—and free of his own leg too! He had to limp home as best he could, and there lay for several days in great pain, with the result that the larders became empty, and he had to live on frogs and weasels—anything, in fact, that he could catch in his burrow.

So, now, if any of you come across a three-legged fox, you will know why it is; and if you happen to catch him, don't keep him, for he is grown up, and grown-up foxes never tame.

AN EGYPTIAN SNAKE CHARMER

Every one has heard of snake charmers. There are many of them in India, and not a few in Egypt too. They walked about the streets of Cairo—or used to do so, for I am speaking of a good many years ago—with boxes and baskets, which contained every imaginable kind of reptile. Whenever they came to what seemed a convenient spot for a performance of their art, they would sit down on the ground, and whilst two or three of them beat on tambourines, a couple more would fill their mouths with a herb, smelling rather like mint, and puff out perfumed clouds of smoke on every side.

When these preparations had been duly made, the sacks, boxes, or baskets were opened; the snakes shook themselves, hissing and wriggling, and began to dance a kind of jig, balancing themselves on the lower part of their bodies, in a way which delighted the spectators.

Besides giving these exhibitions, the snake charmers often go to houses, and after poking all round, at last tell the owners that they feel sure there are snakes hiding there. This is quite enough to cause alarm, for, naturally, no one likes to have such fellow-lodgers, and the snake charmer is paid a certain sum for each reptile he may catch, besides being given the snake itself. He pops it into a bag, and in due time it forms part of his *corps de ballet*.

Now the chief snake charmer in Cairo, whose name was Abd-el-Kerim, had for some time been prowling about the French Consulate, peering in at the doors and windows, and shaking his head in a manner which was far from encouraging.

The French Consul just then was a Monsieur Delaporte, and after a time the report reached him that the Consulate was infested by snakes.

Now, in the course of business, M. Delaporte had come across a good many centipedes, and a certain number of scorpions, but not even the tiniest little asp; so that he had considerable doubts as to the truth of the snake charmer's story. However, at the wish of some anxious friends who trembled at the dangers he might be running, M. Delaporte at last consented to send for Abd-el-Kerim.

The snake charmer was a man between fifty and sixty years of age, clad in a green turban and black robe—grave and dignified—as became his age and profession.

He saluted Delaporte by crossing his hands over his breast, and bowing low before him. Then he waited to be questioned.

'I have sent for you,' said the Consul, who spoke Arabic like a native, 'because I hear a report that there are several serpents in the house.'

The Arab turned his face to the wind, sniffed it up several times, and answered gravely: 'It is true: there are.'

'Oh, indeed! There are serpents?'

'Yes.' And the snake charmer sniffed again, and added, after a moment:

'I may even say that there are several—six of them at least.'

'You surprise me!' said Delaporte; 'and you will undertake to destroy them?'

'I will call them, and they will come.'

'Do so; I should like to see that.'

'You shall see it.'

So Abd-el-Kerim went out from the Consul's room, where this conversation had been held, and fetched in his three companions from the outer chamber. All four men sat down silently on the floor, and after placing their tambourines between their legs, filled their mouths with herbs and began to puff out sweet-scented clouds of smoke, crying: 'Allah! Allah! Allah!' all the time, while Abd-el-Kerim made a hissing, whistling sort of sound, which was intended to attract the serpents.

This went on for three or four minutes without any apparent result; but at the end of that time Delaporte saw about a score of scorpions crawl down the walls or from under the furniture and wriggle up to Abd-el-Kerim.

The Consul's unbelief was rather staggered by the sight of this strange procession. Some of the scorpions came down the mosquito curtains, some down the window blinds, others down the walls; till the thought of sleeping in such a haunted room was enough to make anyone shudder. But wherever they might come from, the scorpions all gathered round Abd-el Kerim, as sheep round a shepherd, and he picked them up by handfuls, and popped them in a goatskin sack.

'You see?' he asked Delaporte.

'Certainly, I see!—I see scorpions, and a great many scorpions, too; but I don't see any snakes.'

'You will see some,' replied Abd-el-Kerim.

And he began whistling in another key, whilst his companions re-doubled their clouds of smoke and their cries of 'Allah!'

And, true enough, to the extreme surprise of the Consul, in a little time a hissing sound, very much like the one Abd-el-Kerim was making, was heard from the sleeping alcove, and from under his bed M. Delaporte beheld a serpent more than four feet long advancing towards the snake charmer, head erect and unrolling his green coils as he glided along.

Delaporte had no difficulty in recognising the species. It was one of those deadly reptiles which the Arabs call *taboric*, and Europeans *Cobra Capella*.

Abd-el-Kerim seized the snake without ceremony by the throat, and was about to stuff it into his bag, when Delaporte stopped him.

'One moment,' he cried.

'What is it?' asked Abd-el-Kerim.

'That serpent was really in my room?'

'You saw it yourself.'

'Very good. Then, as whatever is found in my room belongs to me, be so good, instead of putting the serpent into your goatskin bag, to place it in this bottle.'

And he held out to Abd-el-Kerim a large, wide-necked glass jar filled with spirits of wine, of which he kept a supply in a cupboard ready for the preservation of some of the curious Nile fish sometimes brought him by the fishermen.

'But ——,' began Abd-el-Kerim.

'There's no *but* in the matter,' said Delaporte. 'The serpent was in my house, consequently it is my property, not to mention that I pay you thirty piastres for it. Take care! If you raise any difficulties in the matter I shall begin to think that you put the creature there beforehand, and that it only came to your call because you had tamed it.'

Abd-el-Kerim saw that resistance was useless, and let the serpent glide from his hands into the jar.

Delaporte had a cork and string ready at hand; the cork was firmly tied down on the jar, and the serpent secured inside it.

'Any more?' asked Delaporte.

'Yes,' said Abd-el-Kerim, who did not choose to own himself beaten, and sure enough, after renewed cries and more clouds of smoke, a second serpent, a little smaller than the first, issued from beneath the chest of drawers, and came to Abd-el-Kerim.

Delaporte seized a second glass jar: 'Good,' said he, 'that will make a pair.'

Abd-el-Kerim drew a long face; but he was caught, and there was nothing for it but to give up the second serpent as he had done the first.

'Any more still?' inquired Delaporte.

'No, not here.'

'Where then?'

The snake charmer turned towards the next room.

'I smell one there,' said he.

The next room was the drawing-room.

'Let us go there, then,' said Delaporte. And taking a glass jar under each arm, he gave two others to his servant to carry, and led the way to the drawing-room.

There *was* one there. This one seemed to be a musical serpent, for he had taken refuge under the piano, and in spite of Abd-el-Kerim's manifest reluctance, this snake also promptly found its way into the jar.

'That is the third,' said Delaporte. 'And now, tell me, where are the rest?'

'There are three in the kitchen,' replied Abd-el-Kerim, rather sadly.

'Very good,' said the Consul; 'that will just make up the half-dozen. Let us go to the kitchen.'

At the first call a serpent crawled from under the water-butt.

Abd-el-Kerim placed it in the fourth jar, with a deep sigh.

'Come, come, courage! I want my half-dozen!' said the Consul cheerfully.

'*Enta tafessed el senaa!*' cried the enraged Arab in reply, which, being translated, means 'Certainly you are a spoil sport.' But it was no use.

The snake charmer had to own himself beaten, and in order to save the last two serpents confessed his tricks.

Then Delaporte took pity on him and gave him forty francs, which Abd-el-Kerim pocketed greedily, but could not help murmuring: 'Four serpents which danced so well! They were worth more than that!'

AN ADVENTURE OF GÉRARD, THE LION HUNTER

THE great interest taken in animals by Alexandre Dumas is well known to all readers of the Animal Story Books, but the stories told in them refer generally to tame or tameable animals. The great novelist, however, was full of interest in every kind of beast, tame or wild, and delighted to hear thrilling stories of hunting adventures, and to write them down afterwards for the benefit of his readers.

He was dining with some friends one evening, when his servant asked to see him, and said: 'They have been waiting for you this half-hour, sir.'

Dumas sprang to his feet, and would have hurried from the room at once, but was stopped by the question:

'Who are waiting for you?'

'Gérard, the lion hunter, and his orderly Amida,' was the answer, as Dumas vanished through the doorway in great haste.

In ten minutes he was at home, and there he found the great hunter, and a few other friends all questioning and listening to him.

Gérard, who was an officer in one of the Algerian Regiments of Spahis, was about thirty years of age, with a quiet, gentle face, and clear blue eyes. Amida was a tall stately Arab, of five or six and twenty, and as he sat in one corner of the library, wrapped in his white burnous, he was a striking and picturesque figure.

After warm greetings, and some talk about general subjects and various travels and mutual friends, Dumas sat down to his writing table, drew a sheet of paper towards him, and taking up a pen, he said : 'Now, my dear Gérard, a hunt, come ; anyone at haphazard from amongst your twenty-five lions—but a really fine lion, you know, not one of those you went to see at the Gardens, and which Amida took for sham lions ; but a great, roaring, magnificent lion of the Atlas.'

Gérard smiled, and turning towards Amida said a few words to him in his own language, as though consulting him on the choice of the story. Amida bent his head in assent. Then Gérard turned to Dumas, and in his calm, gentle voice began his story :

I had killed the lioness on the 19th of July, and from the 19th to the 27th I had searched in vain for the lion. I was in my tent with eight or ten Arabs, some my own men, the rest inhabitants of the settlement where I was. We were talking——

'Of what?'

'Why of lions, of course. When you are on a lion hunt, you naturally talk of nothing but lions. An old Arab was telling me a curious legend, several hundred years old, and of which a young girl of his tribe was the heroine.'

'And a lion the hero ?'

'Yes ; a lion.'

'Oh, pray let us hear the legend too,' cried Dumas.

'Very well, then,' said Gérard. 'Here it is :'—

Many centuries ago, there lived a young girl who was very proud and haughty. Not that she was in any way greater or richer than others. Her father had nothing but his tent, his horse and his gun ; but she was very, very beautiful, and it was her beauty that made her so disdainful.

One day, when she went to the neighbouring forest to cut sticks, she saw a lion coming through the trees.

THE LION FALLS IN LOVE WITH AISSA.

The only weapon she had was the little axe which she used in her wood-cutting; but if she had been armed with a gun, a pistol and a dagger as well, she would have been far too frightened to use them—so majestic, proud and powerful was this lion. Her limbs trembled under her, and she would have screamed aloud for help, but her voice died in her throat. She felt sure the lion was going to make signs to her to follow him, so that he might devour her at his ease, in some favourite spot, for lions are not only greedy but dainty.

'I am quite willing to admit that, my dear Gérard,' broke in Dumas; 'but I did not quite understand one remark you made.'

'Which?'

'You said she was sure the lion was going to make signs to her to follow him?'

'Yes. Well?'

'Ask Amida whether, when a lion meets an Arab, he takes the trouble to carry him off?'

Amida shook his head, and raised his eyes in a way which clearly implied: 'Ah, indeed! he's not such a fool as that.'

Dumas pressed for further particulars, and was told what he did not know before, that lions have magic powers. A lion has only to gaze for a few moments at a man, and he completely fascinates him, and the man has to follow the lion wherever he pleases. This point settled, Gérard went on:

The girl then paused, trembling, and expecting a sign from the lion to follow him, when, to her great surprise, she saw him approach, gently, smiling, after his fashion, and bowing in a polite manner.

She crossed her hands on her breast and said: 'What does my lord desire of his humble servant?'

The lion replied quite clearly, 'Anyone as lovely as you are, Aïssa, is a queen, not a servant.'

Aïssa stared in astonishment at this answer, delighted

by the gentle tones of her formidable acquaintance, and surprised that this strange and splendid lion should know her name.

'Who can have told you what I am called, my lord?' she inquired.

'The breeze which loves you, and which, after playing through your hair, carries its perfume to the roses as it sighs "Aïssa!" The stream which loves you, and which, after bathing your fair feet, waters the moss in my cave as it murmurs "Aïssa!" The bird which, since it heard your voice, has been jealous of you, and died of pique as it cried "Aïssa!"'

The girl blushed with pleasure, and began to arrange her veil, taking great care, however, to do it in such a way that the lion could see her all the better; for whether the flatterer is a lion or a fox, and the one flattered an Arab maiden or a crow, you see the result of flattery is always much the same everywhere, and with every one.

The lion, who had hitherto remained at a little distance, now ventured to draw nearer to the girl, but seeing her begin to tremble again, he asked, in his tenderest and most anxious voice: What is the matter, Aïssa?'

She longed to answer, 'I am afraid of you, my lord,' but did not dare; so said, 'The Touareg tribe is not far off, and I am so afraid of the Touaregs.'

The lion smiled, after the fashion of lions. 'When you are with me,' he said, 'you need fear nothing.'

'But,' replied Aïssa, 'I shall not always have the honour of your company. It is getting late, and my father's tent is some way from here.'

'I will escort you home,' said the lion.

Refusal was impossible, and Aïssa had no choice but to accept. The lion came up close, and held out his head as a support, much as a gentleman might offer a lady his arm; the girl laid her hand on his mane, and, side by side, they set out for the tent of Aïssa's father.

On their way they met gazelles, who started away

scared; hyænas, who crouched down in fear; and terrified men and women, who fell on their knees.

But the lion said to the gazelles 'Do not flee;' to the hyænas 'Do not be afraid,' and to the men and women 'Stand up; for the sake of this young girl, whom I love, I will not harm you.'

And all—men, women and animals—gazed with amazement at the lion and the girl, and asked each other, in

THE LION SAID TO THE GAZELLES 'DO NOT FLEE'

their various tongues, whether this strange pair could be going on a pilgrimage to Mecca to worship at the tomb of Mohammed.

At last Aïssa and her escort drew near the settlement, and when they were only some yards from the tent of Aïssa's father, which was the first as you entered the village, the lion stopped, and with the utmost courtesy asked the young girl's leave to kiss her.

Aïssa bent down her face, and the lion lightly brushed her lips with his.

Then he made a gesture of farewell, and sat down to watch till she should have reached her father's house in safety.

On her way there Aïssa turned two or three times, and each time she saw the lion on the same spot. At length she reached the tent.

'Ah! there you are!' cried her father; 'I have been very uneasy.' The girl smiled. 'I was afraid you might have met with some unlucky adventure.' She smiled still more. 'But here you are, and I see I have been mistaken.'

'So you have, father,' said she: 'for, instead of an unlucky adventure, I have had a very lucky one.'

'And what was that?' asked he.

'I met a lion!'

At these words, seldom as Arabs show their feelings, Aïssa's father turned pale.

'A lion!' he cried, 'and he has not devoured you?'

'On the contrary, he paid me many compliments on my beauty, offered to see me home, and escorted me back.'

The Arab thought his daughter must be taking leave of her senses. 'Impossible,' said he.

'How, impossible?'

The father shook his head. 'Do you wish to make me believe that a lion is capable of such attentions?'

Aïssa smiled again. 'Do you wish to be convinced?' asked she.

'Yes; but how?'

'Come to the door of the tent and you will see him, either seated where I left him, or returning to the forest.'

'Wait till I get my gun,' said the father rising.

'What do you want a gun for?' asked the girl proudly; 'are you not with me?'

And drawing her father by his burnous, she led him

to the opening of the tent. But the lion was no longer to be seen at the place where she had left him. She looked all round but could see nothing of him.

'Bah, you have been dreaming!' said her father, as they went back into the tent.

'Indeed I can assure you that I seem to see him still,' replied Aïssa.

'What was he like?'

'He must have been between four and five feet high, and nearly eight feet long,' replied the girl.

'Well?'

'With a superb mane.'

'Yes?'

'Eyes as bright and yellow as gold.'

'Well?'

'Teeth like ivory, but——' and the girl hesitated.

'But?' repeated her father.

'But,' she resumed in a lower voice, ' he had not a very nice smell.'

She had barely uttered these words when a fearful roar was heard just behind the tent, then a second some five hundred yards off, and a third at about half a mile further still.

Then there was silence. Evidently the lion, who no doubt wished to hear what Aïssa would say about him, had made a circle so as to listen behind the tent, and was now hastening away mortified by what he had overheard.

A month passed by, and Aïssa had almost forgotten her adventure, when one day she was told to go to the forest again and cut sticks. Having got what she needed and bound them together in a faggot, she was about to leave, when she heard a slight noise behind her and turned round.

There was the lion, seated a few paces off and looking at her.

'Good morning, Aïssa,' he said, in a dry tone.

'Good morning, my lord,' replied Aïssa, rather nervously, as she thought of the past. 'Can I do anything for your lordship?'

'You can do me a service.'

'What is it?'

'Come near me.'

The girl drew near trembling inwardly.

'Here I am.'

'Good. Now lift up your axe.'

She obeyed.

'Now strike me with it on the head.'

'But, my lord, you— you can't mean——'

'On the contrary, I *do* mean so.'

'But my lord——'

'Strike!'

'Really, my lord?'

'Will you strike?'

'Oh, yes, my lord,' said Aïssa, more frightened than ever. 'Hard or light?'

'As hard as ever you can.'

'But I shall hurt you!'

'What's that to you?'

'And you *really* wish it?'

'I really do.'

So the girl struck as she was bid, and the axe made a deep cut between the lion's eyes. It is ever since then that lions have that straight furrow in their faces which is particularly noticeable when they frown.

'Thank you, Aïssa,' said the lion, and with three great bounds he vanished into the depth of the forest.

'Dear me!' thought the girl, rather hurt at his disappearance; 'I wonder why he never offered to see me home to-day!'

Of course this second adventure of Aïssa's caused a great deal of excitement, but the most ingenious brain could make no guess as to what might be the intentions of this strange and mysterious lion.

AÏSSA'S FATHER FINDS HER AXE

A month later Aïssa once more returned to the forest. She had barely had time to cut a few sticks when the lion emerged from behind some shrubs; no longer gracious and affectionate as at first, or melancholy as at their second meeting, but looking gloomy and almost threatening. Aïssa longed to turn and flee, but the lion's glance seemed to root her feet to the spot. He approached, and she felt that if she attempted to take a step she should certainly fall down.

'Look at my forehead,' said the lion sternly.

'Let my lord remember that it was only by his express orders that I struck him with my axe.'

'I do remember, and I thank you. That is not what I wish to discuss with you.'

'What does your lordship wish to discuss with me?'

'I wish you to look at my wound.'

'I am looking.'

'How is it going on?'

'Wonderfully well, my lord, it is nearly healed.'

'This proves, Aïssa,' said the lion, 'that wounds given to the body are very different from those inflicted on the feelings. The former heal with time, but the latter never.'

This moral sentence was followed by a sharp cry and then complete silence.

Three days later Aïssa's father, searching everywhere for his daughter, found her axe. But of Aïssa herself there was no trace, nor was anything ever heard of her again.

The Arab had barely concluded the legend (said Gérard) when a well-known sound sent a thrill through us all. It was the roar of a lion, probably of the one I had been seeking the last eight or ten days. I sprang at my gun, Amida seized his, and we both hurried towards the spot from which the sound came. It seemed to be more than a mile off. We counted three roars; then the lion ceased, and we marched on towards him.

When we had walked half a mile or so we heard the shouts of men and barking of dogs. We quickened our pace and fell in with a troop of armed men leading a number of dogs of all kinds. The lion had passed that way. He had entered the settlement next to ours, had scaled the enclosure where the flock was kept, and had carried off a sheep. He had secured his dinner; and that was why he had not roared again.

This was hardly the moment in which to attack him; lions do not like being disturbed at their meals. So I begged the Arabs to follow up the track—always an easy matter when a sheep is the victim—and I returned to my tent.

'But why is it easier to track a lion when he carries off a sheep than when he takes some other animal?' asked Dumas.

Gérard smiled. 'That is another story,' said he, 'and if you want to hear it, here it is:'—

One day a lion was talking to the Marabout Sidi-Moussa. Now if the lion is the most powerful of beasts, the Marabout is the most holy of dervishes. So the two were conversing very much on an equality.

'You are very strong,' said the Marabout to the lion.

'Very,' replied the lion.

'And what do you consider the measure of your strength to be?'

'My strength is as the strength of forty horses.'

'Then you can seize a bullock, throw it over your shoulder, and carry it off?' asked the Marabout.

'By the aid of Allah, I can,' said the lion.

'Or a horse, I suppose?'

'By the aid of Allah, I can carry off a horse as easily as a bullock.'

'Or a wild boar?'

'By the help of Allah, I should do with the wild boar as with the horse.'

'And a sheep?'

The lion began to laugh; 'I should think so!' said he.

But the first time the lion captured a sheep he was much surprised to find that he could *not* throw it over his shoulder, as he did with far larger and heavier animals, but had to drag it along the ground. This was the result of his proud boasting, and of forgetting to say,

THE LION LAUGHS AT THE MARABOUT'S QUESTION

as he did about the larger animals: 'By the aid of Allah!'

Ever since then the lions have been obliged to drag any sheep they may capture along the ground, leaving a track after them.

So you see why I felt sure of being able to track my game later on. Well, I had hardly regained my tent when the owner of the sheep arrived, hot and panting, and told me that he had followed the traces of the lion for a mile and a half, but had been unable to go further. However, all his information was very precise, and I was able to give orders to my two beaters, who, luckily, were experienced men, for a track is far more difficult to follow up in summer than in winter.

They were both Arabs, from thirty to thirty-five years of age, strong, hardy, and cunning—true sons of the desert.

One was called Bilkassem, and the other Amar Ben-Sarah.

They divided the work between them, Bilkassem taking the animal from the time he left the settlement, and Amar Ben-Sarah from the point where the owner of the sheep had lost the track.

After a search of nearly two miles, Bilkassem found the skin of the sheep—for the lion is a dainty animal, and does not eat hides; and, on reaching the neighbouring well, Bilkassem found a mark left by Amar Ben-Sarah. It was needless for him to go any further. His comrade was on the track, and he knew there was not much chance of its being lost. So Bilkassem returned to the tent and brought me his report.

Meantime Ben-Sarah followed the lion.

Towards mid-day Amar Ben-Sarah returned too. The lion had retired into its lair. The Arab had described a circle of a thousand paces round his den, and thus made sure of finding the exact spot. It was nearly 4,000 yards off.

My mind was made up, in all probability we should meet that very day.

The day wore on. I felt nervous and excited, and could neither eat, read, nor occupy myself with anything, in my feverish impatience, and shortly before sunset I set out. It is the time when any natives who may happen to have a lion in their neighbourhood invariably stop at home. From the first moment of the short twilight till the following day, any Arab who has heard that warning roar feels the greatest reluctance to put a foot outside his tent. But the very reason which kept them safely indoors determined me to choose this particular hour, for this is the time when the lion awakens from his mid-day sleep and starts out in search of prey.

When I reached the place marked by Amar Ben-Sarah I found I still had a quarter of an hour's daylight, and might study the landscape.

It was the entrance to a mountain gorge. The slopes on either side and the bottom of the gorge itself were thickly wooded, the trees interspersed here and there with bare rock, which stood out like gigantic bones, and were still burning after the heat of the day.

We plunged into the gorge, Ben-Sarah acting as guide. Behind him he dragged a goat, who was to serve as a decoy for the lion.

About fifty paces from the lion's lair there was a clearing, which I chose as my point of vantage. Amar cut down a sapling, sharpened one end, and planted it firmly in the middle of the clearing. Then he tied the goat to it, leaving its rope a couple of yards long.

As he was completing his operations we heard a loud and prolonged yawn at no great distance. It was the lion, only half awake as yet, but who was looking at us, and who yawned as he looked.

The bleatings of the goat had wakened him. He was quietly sitting at the foot of a rock and deliberately licking his thick lips, looking all the time full of the most magnificent contempt for us.

I hastened to order my men back, and they were not sorry to take up a position some two or three hundred yards behind me. Amida alone insisted on remaining close by me.

I carefully examined the spot. A ravine separated me

from the lion. The clearing might be forty-five paces round, consequently fifteen paces across.

I was alone, and had to choose my place. I took up a position at the very edge of the wood, so that the goat was between the lion and me—the goat was seven or eight paces from me, the lion about sixty.

Whilst I had been making my little inspection the lion had disappeared; there was evidently no time to be lost in preparing to receive him, as he might fall upon me at any moment. An oak tree offered the support I always look for on these occasions. I cut off the small boughs which might have hindered my movements, and sat down with my back against the trunk. I was hardly seated before the signs of agitation shown by the goat told me plainly that something was going on close to us. The goat dragged at his cord with all his might towards me, but kept his eyes fixed on the opposite side.

I understood that the lion had taken a roundabout path to reach us, and was now approaching, following, as he did so, the fold of the ravine.

I was not mistaken. At the end of ten minutes I saw his huge head appear at the top of the ravine which had at first divided us, then his shoulders, and then his whole body. He walked slowly, not yet fully awake, and with his eyes half closed, for the lion is a great sleeper and very lazy.

Having reached the top he found himself about seven paces from the goat and fifteen from me. I remained settled where I was, and took aim at him right between the eyes. For a moment I felt tempted to pull the trigger, but the fascination of watching the superb creature and noting the movements and ways of my formidable antagonist kept me motionless. For some moments I enjoyed such an interview as few men can boast of. I felt I deserved it, for it was two years since I had been actually face to face with a lion, and this was one of the finest and largest I had ever seen. At the end of a few minutes he

THE LION APPEARS AT THE TOP OF THE RAVINE

crouched down perfectly flat on the ground, then he crossed his paws in the front of him and pillowed his head upon them. His eye was fixed on me, and his glance never wavered from mine for an instant. He seemed to be wondering what this man could be doing in his kingdom without even recognising his royalty. Five minutes more passed. In the position he had taken up nothing would have been easier for me than to have killed him.

All of a sudden he rose, and began to be agitated, making a couple of steps forward, then one or two backwards—to the right, to the left—and moving his tail like a young cat who is getting angry.

No doubt he could not understand this goat with its cord or this man who kept watching him, but his instinct told him there was some trap.

Meantime I sat quite still, the gun at my shoulder and my finger on the trigger, following every movement with my eye. One spring, and I should be between his claws. His anxiety increased every moment, and almost infected me. His tail lashed against his sides, his movements were more rapid and his eye kindled.

To hesitate longer would be suicidal. I seized the moment when he turned his left flank towards me, took a steady aim and fired.

The lion staggered on his legs and uttered a frightful roar, but did not fall.

I fired my second shot. Then, without looking, for I was sure I had hit him, I threw down my first gun and seized the second which was lying ready loaded beside me. When I turned round again the lion had disappeared. I remained motionless, fearing a surprise, and looking round on all sides for a hidden foe.

I heard the lion roar. He had fled into the bed of the ravine, and was hurrying back to his lair.

I waited a few minutes more, or perhaps they were only seconds, for one does not measure time accurately in such circumstances.

Then, hearing nothing, I rose cautiously and went to inspect the spot where the lion had received my two shots.

The goat was panting on the ground, terrified, but otherwise unhurt.

I soon realised that the lion had been hit by both my balls, and they had pierced him right through. Every hunter knows that an animal can go further with a wound right through the body than if the ball is lodged in its inside. I set off on the track. It was not difficult to follow.

As I supposed, he had regained his lair. At this moment I saw the heads of Amida, Amar Ben-Sarah, and Bilkassem appear at the top of the ravine. They approached with caution, not knowing whether I was dead or alive, and prepared to fire. When they saw me they shouted with joy and ran to join me. They wanted to start at once in pursuit of the lion, but I held them back; for, in my opinion, the lion had been dangerously, probably mortally, wounded, but the heart had not been touched. He was still full of strength, and his last struggles would be terrible.

As we were discussing this, eight or ten more men, armed with guns, joined us. They had heard my two shots, and, like Amida, Bilkassem, and Amar Ben-Sarah, ran to see what had happened.

Their first cry was 'Let us follow him!'

I assured them they would run great danger. But no; 'Stay there,' said they, 'and we'll bring him to you dead.'

It was useless to repeat that the lion, in my opinion, was still very much alive indeed; they insisted on entering the wood.

Finding that nothing would turn them from their project, I determined to go with them. But I took my precautions. I reloaded my favourite gun, gave one to Ben-Sarah and another to Amida, and, thus prepared, I entered the wood on the track of the lion.

It was almost dark; the wood was thick with shrubs and undergrowth, and one had almost to crawl along.

My three Arabs followed me, and the men from the settlement came behind them. It took us nearly a quarter of an hour to walk fifty steps, and even that we did with much difficulty.

After fifty steps more it was quite dark, and we had lost the track.

There was a clearing close at hand, and we made for it so as to reconnoitre.

Whilst we were scattered about in the clearing, trying to make out some vestige of the track, either by accident or by awkwardness a gun suddenly went off.

Instantly a hoarse roar was heard, and the lion fell amongst us as though he had literally dropped from the clouds.

There was an instant of intense terror. Every gun except mine went off at the same moment, and it was only a wonder that we did not all kill each other. Needless to say that not a ball touched the lion.

Through the fire and smoke I saw all the men round me except Amar Ben-Sarah. Then from the other side of the clearing I heard a piercing cry.

I ran towards the spot from where the sound came, and in the dusk only saw the man and the lion when I was actually upon them.

Amar Ben-Sarah was lying on the ground, and the lion standing over him.

I felt giddy, and thought my legs were going to bend under me, but the weakness passed like a flash of lightning.

The lion, seeing the muzzle of my gun so near his head, turned to me with a savage look. In another second he would have been upon me; but I was too near to miss the fatal spot. I pulled the trigger, he staggered a few paces to one side, and then dropped down dead beside the man he was about to kill.

PUMAS AND JAGUARS IN SOUTH AMERICA

No one can have read Captain Mayne Reid's stories about America without being struck by the part played in them by an animal called the 'painter,' which is of a tawny colour, with a black stripe down its back. Now the 'painter' is really the panther, and the panther is the creature that we call the puma, which, next to the jaguar, is the biggest of all the American cats, and has a wider range than any other mammal. The puma is to be met with in British Columbia, or in the Adirondack mountains not far from New York State; it is to be seen in the hot unhealthy swamps that lie along the northern shores of the Gulf of Mexico; it lies in wait for its prey in the river forests of the Amazon and the Orinoco; it tracks the wild and cunning huanaco ten thousand feet high on the Andes, and it is the dreaded enemy of colts and sheep on the cattle runs of the Argentine Republic. With wonderful skill it makes the best of circumstances; if horses, its favourite food, are not to be had, it puts up with ostriches; if it happens to live in Mexico, or Arizona, it makes its dinner off wild turkeys; further north still, the puma will be content with porcupines or even snails, while if its chosen haunts along the river banks of the Amazon or the Orinoco are overwhelmed by a sudden inundation, it takes to the trees and feasts upon monkeys.

As sometimes occurs in families, the puma has a particular hatred for its cousin the jaguar, and seldom indeed does it fail to get the better in any fight. It also has a violent dislike to dogs, and in South America can never see one without flying out to attack it, while the grizzly bear is its deadly foe. But, on the other hand, in the great continent of South America it shows its best qualities. It loves man, and even when attacked by him will not defend itself, while in puma-haunted districts children may even sleep all night alone, without fear of harm. And not only children, for travellers tell us a puma has never been known to attack a sleeping man.

It is a great pity that pumas are so fond of killing tame and useful beasts, as they have many delightful qualities as pets. Pumas are very playful, and very affectionate and gentle to people and children; but they are rapidly being hunted down, as farmers find it quite impossible to keep any cattle in their neighbourhood. Between their courage and their wonderful powers of jumping, no animals are safe from them. Some witnesses have declared that pumas have been seen, when pursued by dogs, to spring a clear twenty feet into the air for shelter in a tree, while another leap of forty feet was measured on the ground. In Patagonia, a farmer who had suffered much from a puma's appetite shut all his sheep into a huge fold, surrounded by a wooden paling fifteen feet high. The only entrance was by a six-foot gate, and, to make all secure, men and dogs were told off to watch. But the puma was too clever for them all! He seized his chance when any clouds came up to make the darkness thicker, and every morning one sheep at least was found with a dislocated neck, and its breast eaten, for this is the way a puma always kills its prey, and, except when very hungry, it never eats the whole carcass. One night, the naturalist[1] who tells the story was passing by the gate, when the robber sprang right over his head, but it was too dark to give chase, so the puma got away safely. Afterwards, it

[1] Hudson.

was found that it had been in the habit of hiding till dark in the pen with some calves, which it never tried to touch, as it knew it was sure of the sheep.

In many places in Patagonia, where horses are bred, farmers have been obliged to turn their attention to something else, as the colts invariably fall victims to the pumas. They will lie patiently in wait for them to pass, and, never caring for the man, or men, who may be bringing the drove back from pasture, will spring out from behind a bush right on the back of the colt, place one paw on its head, and the other on its bosom, and bring the head back with a jerk. Then, before the driver has had time to come up, the puma is deep in the bushes again.

There is nothing mean about a puma; it is all the same to this great big cat whether the beast it is hunting is large or small, fierce or tame. It will trot, or rather bound, after a peccary, a jaguar, or a grizzly bear, quite as cheerfully as if it were stalking a colt or a sheep. Only one animal has been known to get the better of a puma, and that is the last you would ever expect — a donkey. It is the fable of the hare and the tortoise over again. The puma may jump on his back as much as it likes, the donkey puts down his head, so that the puma cannot seize his neck, and kicks so hard that the puma is at last shaken off; or if that does not do, the donkey takes to bucking, and anybody who has ridden much knows very well what the end of bucking is likely to be.

But when pumas can be kept away from all other beasts, and be seen only with man, or with each other, what charming and graceful creatures they show themselves! Fancy watching pumas chasing butterflies for the pure fun of it; or playing with their babies as if they were so many kittens, rolling them over and stretching out their tails for the little ones to catch, or having a game of hide and seek behind the rocks and bushes. It seems almost absurd to think that a puma could ever want to hurt any living thing—and if you had not seen a cat's eyes when it looks at a bird, you might say the same about *him*!

But many are the stories told in South America of the attachment of the puma to man, and the kindness it has shown him. One day, a band of men went out to hunt, and scattered in search of game all over the plains or pampas. In the evening, when they all assembled to ride home, one of the number was missing; but on reaching the farm, his horse was found quietly standing outside his stable. It was too late and dark to do anything that night, but at dawn next morning the rest set forth, and after some hours they found their missing comrade, lying on some ground, with his legs broken. The poor man had spent a terrible night, for the voices of jaguars were often heard in the distance, and most likely would have come a good deal closer, had it not been for a puma, who had never ceased walking about as if to guard him. When the jaguar's voice became louder than usual, the puma crawled silently and noiselessly away, and sounds of battle came through the darkness. No more was known of that jaguar.

There is an old legend which is to be found in every history of the Spanish settlers in South America, that seems almost like one of the stories of the early martyrs. In the year 1536, says Ruy Diaz de Guzman, the Spanish settlers in the town of Buenos Ayres were closely besieged by Indians, and, after suffering frightful hardships from hunger and thirst and sickness, eighteen hundred of the unfortunate people died, and were buried, by the six hundred that were left, just outside the wooden palisade that defended them from their enemies. The graves were dug hardly below the surface of the ground, for the diggers looked up with fear between the turning of every sod to see if the Indians were approaching, and the smell of the dead bodies soon attracted swarms of wild beasts from the country round, so that on every side the Spaniards were beset with dangers.

At last, most of the few who were left declared they could bear this state of things no longer. It was a choice of evils, and they made up their minds that they would prefer to fall into the power of beasts rather than of men. So, when the darkness had fallen, a little company crept out from the palisade, and stole away to the woods.

How they fared we are not told; but one girl, called Maldonada, after wandering about till dawn, fell in with some Indians, who carried her off to their village in the heart of the forest, and treated her with great kindness.

Some months later, Ruiz, the deputy-governor of Buenos Ayres, heard where she was, and—being by this time free from his enemies—sent to the friendly tribe to beg of them to give Maldonada up to him. When the poor girl was brought back to the city she found that it was only to be accused as a traitor to her own people, and to be condemned to be fastened to a tree in the forest, so that savage beasts might devour her.

So Maldonada, who had passed unhurt amidst the hungry animals, whose midnight wars she had heard when flying from the besieged city, was now to be delivered over to a fate from which no escape was possible. *How* a girl living quietly in an Indian village *could* have betrayed her people, Señor Ruiz did not say, and it is not clear why he was so anxious for her destruction; but sentence was given, and the soldiers called in. They led Maldonada three miles into the heart of the forest, and there tying her tight to a tree, according to their orders, left her to her death.

For two nights and a day no one troubled their heads about her; either she had no friends, or they were poor people who were powerless against the governor; but on the third day, soldiers were again sent out, to collect her bones. To their immense surprise, they found Maldonada quite unhurt, but very hungry, and awaiting, as bravely as she could, the death that could not be far off, whether it came to her by starvation, or by the jaws of wild beasts. During the terrible hours she had spent there savage

MALDONADA GUARDED BY THE PUMA

creatures of all sorts had tried to get at her, but had been driven off by a puma, which had stood by her side, and defended her from every enemy, and, according to one writer, dead and dying jaguars were scattered round.

When the soldiers came up, the puma retreated to a little distance, fearing two or three men more than any number of wild beasts. But when, moved to pity by Maldonada's wonderful deliverance, they unbound the ropes that fastened her, the puma drew near again, and jumped about her, and rubbed its head on her shoulder, and showed how pleased it felt that all its battles had not been fought in vain. 'And in this way,' ends up the old chronicler, 'she who had been offered up to wild beasts became free. I knew her well, and think that, instead of being named "Unlucky," she should rather have been called "Lucky," and the things that happened to her show plainly that the punishment meted out to her had in no manner been deserved.'

A jaguar is a very near relation to a puma, though they are deadly foes; and it is the biggest of all the cat tribe throughout the continent of America, measuring over six feet from nose to tail. Its skin is yellow, spotted with black, and, like the puma, it is a very clever climber, and can manage quite well to dine off the monkeys that live up on the trees, if the solid ground is flooded. It is, perhaps, the fiercest of all the wild animals of South America, and it is certainly one of the noisiest. The puma goes silently about its business, but the jaguar is always shrieking and screaming, so its prey has plenty of warning, and can often get safe out of the way.

The jaguar is found all through America, from Texas to Patagonia—an immense tract of country, that of course contains a great many different climates, to which it has to adapt its food and habits. In the forests which border the Amazon, and some of the huge rivers of Brazil, they make their lairs along the banks, or in the reedy shores of the lakes. Here they feast—for a change, or when nothing else is to be had—on fish, eggs, and even turtles, which they scoop neatly out of their shells with a paw. Sometimes they inhabit the islands scattered about the great streams; but when the rivers suddenly rise, and their homes are flooded, and no food is to be had, they swim on shore in search of it, and it is at these times the jaguar becomes unusually dangerous, for, as a rule, it never attacks man first. On one occasion, a half-starved jaguar hid itself in a church at Santa Fé, and as the priest entered to celebrate mass, it sprang out and gnawed the poor man, till there was hardly a scrap of him left to tell the tale. Then the murderer stole stealthily back to its hiding-place, with its appetite still keen, waiting till the second priest should come in and fall a victim, exactly as the first had done. And even two priests would not have made a meal for this hungry creature, but that fortunately the third priest, whose ears were quick, heard the sound of crunching through the open door, and stopped outside in time.

He rushed back and collected some men, but no one could be found rash or daring enough to advance into the church in order to shoot the monster. It was found that the only safe way to get at it was to go up on the roof of the church, and to lift off a part, so as to take aim from a safe distance.

When goaded by hunger, jaguars will eat tame cattle and horses; but they much prefer wild game, which they kill in the same way as the puma, by dislocating the neck. If they are disturbed during a meal they will hardly ever return to the half-eaten body, but begin a fresh hunt—and in the level pampas of Argentina and Patagonia, game is very plentiful and easily seen. In the southern parts of Brazil man-eating jaguars are not at all uncommon, and one will be heard (or seen) tracking a party during a whole day, stopping when they stop, and moving when they move.

THE JAGUAR BESIEGED BY PECCARIES

In his 'Pioneering in South Brazil,' Mr. Bigg-Wither tells a curious story of a fight between a jaguar and a herd of wild pigs, witnessed by some friends of his who were exploring the country. One evening the two men had come in very tired after a long day's work, during which they had eaten nothing but fruit and honey, and set up their camp in a belt of forest between two rivers. They were sitting round their fire feeling very hungry and longing for a good meal, when suddenly a great noise of grunting and squeaking close by betrayed the presence of pigs, and the men seized their guns and pricked up their ears, thinking that here at last was their chance of a dinner.

Going cautiously in the direction of the sound, they came upon a clearing. In the midst of the clearing was an anthill, about five feet high, and on the top of the anthill stood a large jaguar. Round the foot of the anthill were a herd of fifty or sixty wild pigs, grunting, squeaking, and bustling noisily about, but not knowing how to get at the jaguar, who stood balancing himself uneasily on the crest of the anthill, with his four feet well together, and his tail high in the air, out of harm's way.

But it was plain to the two men who were watching that this state of things could not last long, and, indeed, very soon, either from forgetfulness or from laziness, the jaguar allowed his tail to drop a little. In an instant it was seized by a smart young pig, and the jaguar dragged right down among his enemies, who closed in a dense mass round him. In spite of the immense odds against him, the animal fought well and pluckily. Two or three times he actually struggled to his feet, and struck out fiercely with his paws; but the battle was against him, and little by little the noise began to cease.

Then the pigs slowly dispersed, and sauntered off by ones and twos and threes, some in this direction and some in that. When they were all out of sight the men came out from behind their tree, and walked quickly to the battlefield, where fourteen pigs lay dead or dying, but nothing was to be seen of the jaguar. Where *could* he be? was the question they asked each other, and the riddle was only guessed when one of the men, who was a Portuguese, picked up a bit of his skin. Whether he had been torn and eaten on the spot, or whether he had been carried off piecemeal to be enjoyed at home, was never known. Anyhow the men took one of the dead pigs back to the camp, and cooked it for supper.

Belt, the Naturalist, when travelling in Nicaragua, had some interesting encounters with jaguars, which might have ended badly for him. One day he had gone in search of some small birds that feed on foraging ants, and hearing their notes, he tied his mule to a tree, and went in search of them, as he was very anxious to obtain a specimen. He had only with him a gun loaded with very small shot, and holding this he pushed through the bushes to the thicket from which the birds' song came.

But birds are restless creatures, and these must have fluttered from tree to tree, so that Belt had wandered a good way from the path, and had reached a space where the brushwood was thin, and the trees large and tall, when he heard a sound between a cough and a growl from the bushes on his left. He thought it was a tapir, and ran quickly towards it, as he knew that, with such small shot, he would have to be very close before he fired. Then, just in front of him, the bushes swayed, and out came a huge jaguar, lashing its tail and roaring with anger.

It was not easy to tell what had excited it so, for it had not seen Belt, and there was no animal in sight; but it crossed the clearing twenty yards in front of Belt and dashed on. The Naturalist was quite unarmed, except for his one little gun, and knelt down to steady his aim, in case he might have to fire at close quarters. The slight rustling attracted the attention of the jaguar, who paused for a moment, and then turned round. It lowered its head and stretched itself out, and Belt made ready to receive its spring, but the jaguar altered its mind at the last, and bounded off into the forest. It was much the best thing for everybody; but Belt never ceased being sorry that he had not fired, although, if he had, he would most likely never have come home to tell the tale.

In this part of the world, too, jaguars have a peculiar way of killing their prey, which certainly spares the victim any pangs of terror. A jaguar will sit quietly on a tree till a herd of wild pigs come by, and then, choosing out a nice fat one, drops straight on its back as it passes underneath, dislocates its neck with a jerk of its paw, and is up the tree again before the rest of the herd know what has happened. When they have disappeared, leaving their dead comrade behind them, the jaguar jumps down and eats him for dinner.

MATHURIN AND MATHURINE

IN the small village of Saint Jean, near Carcassonne, there dwelt a young man named Mathurin, who made his living by selling milk. This he always carried on his shoulder when he went his rounds, in a large earthen jar, but one unlucky day when he was going over a piece of rough rocky ground, overgrown with gorse and heather, his foot slipped, and his jar fell on a stone, and was broken to atoms.

Close to where the accident happened the rocks formed a little hollow, into which the milk flowed, and soon formed a small white lake. There was no use trying to pour it back again, for the jar was too badly broken for that, so the young man returned as fast as he could to Saint Jean to get some more milk for his customers.

This time he took care to get a stronger pot, and to hold it more firmly on his shoulder; and then he made haste back along the path he had come, for it was getting late, and everybody would be thinking about breakfast.

On reaching the place where he had slipped and fallen, he found that a splendid adder had taken advantage of his misfortunes and was lapping up the pool of milk with the utmost enjoyment. As he came near, the adder turned and hissed, and showed quite plainly that she did not intend to allow anybody to interfere with the piece of good luck which had fallen in her way. The young milkman understood the hint, and was, besides, in a hurry, so he passed on quickly, and left the adder to finish her breakfast.

Still, he felt rather curious to know if she could possibly drink up so much milk, and when he had served all his customers he took the trouble to come back the same way to see what had become of the adder. He found her stretched out on the rock, quite drunk with milk, and being a young man of kind heart, he left her to have her sleep out, instead of killing her, as most people would have done in his place.

MATHURIN AND MATHURINE

The next morning, when he passed by the dry little hollow which, the day before, had been a milky lake, he thought of the adder, and how dull it must feel after its delightful meal of yesterday. So he looked at his jar and then at the hollow, and then at the jar again. Finally he stooped down, and poured out a little milk, and walked quickly away. When he had gone a few steps, he glanced round cautiously, and saw the adder in the act of gliding out from under a bush and making straight for the milk.

This time there was not enough to make it drunk, for when Mathurin came back, an hour or two later, the adder had disappeared.

The following day he looked about to see if the adder was anywhere in the neighbourhood, and detected two bright eyes and a small flat head, watching him from under a bush. He called it by the first name that occurred to him, which was 'Mathurine,' the feminine of his own; the adder seemed to listen. Then he poured out some milk, and called it again. The adder seemed to understand, and came about a yard nearer, then stopped doubtfully.

The young man did not want to frighten her, so he moved to a little distance, but not without seeing his new friend busy over the milk he had poured out. He did not go near her again, but called gently, 'Mathurine! Mathurine! Mathurine!' and each time the adder lifted her head and looked at him.

From that day he never passed the place without calling 'Mathurine!' and at every call the adder hastened more quickly to answer it, till she soon became quite tame, and recognised not only the young man's voice but the sound of his footsteps.

The friendship between this odd pair lasted for a year. Every day during that year Mathurin poured out a glass of milk for Mathurine, and every day Mathurine was on the look out for him, standing on her tail when he appeared, and licking his hand affectionately with her forky tongue.

But at last there came a day when the young man drew the lot of conscription, and had to leave the village where he was born, and join the regiment to which he was appointed.

He bade an affectionate farewell to his little friend, who had grown quite a foot during the last few months, and was now as tall as Mathurin himself when she reared herself to her full height. She quite understood that she was not going to see him for a long time, and overwhelmed Mathurin with caresses; curling about his legs and arms, and rubbing her head against him. Then she glided by his side for part of the way, and only vanished among the bushes at the sound of the bells of the stage coach.

Mathurin was away seven years, from 1793 to 1800— a rather lively time he had—and it was only after the peace of Lunéville that he was set free to return home, with the uniform of a corporal.

His first visit was of course to his mother; then to his sisters, his cousins, and his friends. After that, he thought about the adder. Would she remember him, he wondered, after seven years' absence? He was curious to know.

He put on his old milkman's clothes, so that Mathurine might the more easily recognise him, and went straight to their old meeting-place in the rocks. 'Mathurine! Mathurine!' cried he.

Instantly there was a loud rustling among the leaves, and a snake ten feet long, with gleaming eyes, came wriggling along with amazing quickness and flung herself with a bound upon Mathurin, twining herself tightly round his neck. He tried to free himself from the pressure which threatened to choke him, but could not unloose the closely curled rings; then he attempted to call for help, but his voice died in his throat, and, throwing his hands despairingly in the air, he rolled dead upon the rocks, strangled by the embraces of his friend.[1]

[1] The young reader is requested to correct the mistakes in this exercise of French fancy.—A. L.

JOSEPH: WHOSE PROPER NAME WAS JOSEPHINE

MONSIEUR ALEXANDRE DUMAS, who was so fond of animals, and has given us such a delightful account of 'Pritchard and his ways, was once passing a few months in a palace at Naples.

It was a beautiful palace, with a garden that had been made long ago by a rich Roman noble, and terraces that sloped down to the sea itself. These terraces and gardens were filled with fine trees and covered with flowers, and on their walls and stones there basked in the sunshine, thousands of grey and golden lizards.

Now anyone that has ever watched the behaviour of lizards for long together, knows what strange little creatures they are. How quick, and yet how still; how shy, and yet how readily tamed; how unnoticeable amidst the grey rocks and stones, yet how easily detected by their bright glittering eyes.

Amongst all the lizards that made their homes in the gardens of M. Dumas' palace, there was one which seemed as if it had been charged by all its relations to prove to M. Dumas and his guest, M. Goujon, the truth of the proverb, 'the lizard is the friend of man.' This particular lizard was a very bold little person, and very fond of flies, which it would even come to seek by the windows of M. Goujon's room, opening on to the terrace.

Like M. Dumas, M. Goujon loved beasts, and he thought he would try to tame his visitor, and at the end of three days he had succeeded so well, that the lizard was not afraid to come near him. A week later he tried the experiment of offering the lizard, to whom he had given the name of Joseph, a spoonful of tea from his cup, and, rather to his surprise, Joseph seemed quite to enjoy it!

The two always met in the very early mornings before anyone else was up, but at whatever hour M. Goujon might choose to come out on the terrace, Joseph was sure to be there before him, stretched comfortably out in a warm sunny place, with her eyes fixed on the door where M. Goujon would presently appear.

Ten minutes after this event, a page boy brought Goujon his morning tea, and Joseph, who knew the boy quite well by sight, raised her head and flicked the end of her tail with joy at the sight of the tray. She never moved her gaze from Goujon, who poured himself slowly out a cup, and put in plenty of sugar. Then he took a spoonful of the tea, tasted it, as a careful nurse tastes a baby's milk to make sure it is not too hot, and held out the spoon to Joseph, who lapped it delicately with her thin black tongue till she had finished every drop. She never allowed anything to disturb her during this occupation, except a sudden noise, or a face she did not know.

Little by little Joseph soon grew accustomed to the people of the house, and paid no heed to them. She would even take sugar from our hands, though seldom without hesitation, as she always remained constant to Goujon.

One day Joseph was missing from her usual place on the terrace, and M. Goujon had to drink his tea alone. The whole house grieved over her loss, for 'the palace was dull,' observes M. Dumas, 'and we had made no friends except herself. But there is no sorrow so great

that time cannot heal it ; and, as Claudius King of Denmark
said to Hamlet on a similar occasion :

> .... Your father lost a father ;
> That father lost, lost his ; and the survivor bound,
> In filial obligation, for some term
> To do obsequious sorrow : but to perséver
> In obstinate condolement, is a course
> Of impious stubbornness ; 'tis unmanly grief.

Hamlet refused to listen to this advice ; but, as M.
Dumas afterwards said, in telling the story, 'We were
wiser than Hamlet. Besides, after all, Joseph was not
the father of any of us. If she was anything, she was
Goujon's adopted child.'

However, all missed her, and for two or three days
she was the subject of all our conversations. Then
her name was heard more seldom, and at last it dropped
out of our talk altogether. Only Goujon would every
now and then lean over the parapet, and call softly
for 'Joseph,' and even he seemed to do this now more
as a matter of duty, than from the idea that it was of any
use.

Things went on in this way for about three weeks,
when, early one morning, at the hour when Goujon was
in the habit of drinking his cup of tea, I heard cries of
joy proceeding from the terrace. I ran to see what had
happened, and found Goujon wild with delight at the re-
appearance of Joseph (or Josephine as she ought properly
to have been called), who was basking in the sun with
two tiny little lizards about as long as needles and as
thick as quill pens, lying beside her.

She stayed with us till the middle of November, and
then vanished as suddenly as before. Nothing was seen
of her during the cold days of the winter, but at the
beginning of March, when the sun was growing strong
again, we noticed one morning a lizard lying on the wall
of the balcony, staring hard at us.

JOSEPH'S BREAKFAST

'Look there,' I remarked to Goujon. 'One would almost say that was Joseph,' for we never could remember to call her 'Josephine,' and, if we did, she paid no sort of attention.

Goujon's eyes followed mine. 'Joseph, Joseph!' cried he, and Joseph came running without a moment's hesitation, to the astonishment of the two small lizards, who stayed behind and watched, with a shudder of horror, their mother crawl up Goujon's shoulder.

From this day the friendship between Goujon and Joseph became as strong as ever, and when we left our palace the only person we were sorry to leave was the amiable Joseph.

Does she ever think of us now, I wonder, even of Goujon?

THE HOMES OF THE VIZCACHAS

ONE of the most curious and interesting of all the dwellers in the pampas of South America is a little fat creature, rather like a large guinea-pig, found from the Rio Negro to the Uruguay, and called the Vizcacha. It is nearly related to the Chinchilla, but does not enjoy mountain life or solitude, and, indeed, prefers to live in a settlement with twenty or thirty companions.

Like the beaver, the vizcacha is a great builder, and his houses are always made on the same plan. He first of all chooses a level spot, where the soil is neither sand nor gravel, and then digs deep trenches or passages which lead into the inner apartments, the front doors being very large and handsome, often as much as four or five feet wide. At the end is a large round room, and the whole dwelling is in the form of a Y.

Of course, during the process of building, a great deal of soil has to be thrown out, and the vizcacha, who is very neat and thorough about all he does, erects this into a mound, which serves as a protection to the burrow and prevents it being trampled under foot by the passing cattle, or being washed away by the heavy rains, as often happens to the homes of armadillos and other animals. On the sides of this mound burrowing owls make their nests, and various small birds are to be found that exist (as far as is known) in no other place, while foxes and weasels find it quite a pleasant residence. These vizcacha burrows, or vizcacheras, as they are called, often cover as much as two hundred square feet of ground, and are so numerous in Patagonia that you can hardly ride half a mile without coming on one at any rate. The villages go on in the same place for generations, except that every now and then, when the dwelling is getting uncomfortably crowded, a vizcacha of unusual energy will look out for a suitable spot fifty or sixty yards from his old home, and form a new burrow; his lazy companions, however, taking care not to join him till all is ready, when they drop in by accident one by one.

When once the vizcachera is built, in nice soft ground, and its park (about half an acre in extent) of smoothly cropped grass is properly laid out, the vizcachas show themselves to be people of regular habits.

In winter it is their custom to stay in their burrows till dark, but in summer they come out before sunset, to take advantage of the evening air. First one of the elders will appear and sit quietly on the mound, and then, gradually, the doorways are filled with loungers, the males standing upright, and the females, smaller and livelier and lighter than their masters, sitting on their haunches. Like their two-legged sisters, they become eagerly curious at the sight of any passer-by, and make strange noises. If he approaches, they dash quickly into their burrows; but often their sharp eyes and little noses may be seen peering round the corner, longing for another look.

All vizcachas are very careful about their fur, and spend much time combing it out smooth with their paws. They are very sociable, but do not consider it good manners to enter each other's houses; visits are paid at the entrance, and even when pursued, a vizcacha will hardly ever seek refuge across his neighbour's threshold. They have no idea of self-control, and any sudden noise occurring when they are feeding gives rise to a perfect babel of cries and screams. A vizcacha has a great variety of notes, and can make himself heard at a great distance. He also has a very odd trick of stopping in the middle of his dinner to utter loud shrieks, and at night he never seems to stop talking, as dwellers on the pampas know too well.

Vizcachas are hardy little creatures, who can do without water as long as they can get green food. But in dry summers, when nothing is to be had but withered grass or dry thistle-stalks, they are forced to drink when they can. They are very busy and energetic, and when once their house is ready, time seems to hang heavy on their hands—at least, that is perhaps the reason why they are so careful to leave nothing lying about, but drag every kind of refuse to the mouth of the burrow and pile it up in a mass. This trick is so well known, that, in the pampas, when any article is missing, it is at once looked for in a vizcachera, just as we should search a magpie's nest, and it is on record that a man's watch was once discovered there. The little animals show a sense of fun, too, which must make them amusing to watch, especially in their dealings with dogs. Except when they are feeding, when anything upsets them, the appearance of a dog produces no effect on the nerves of a vizcacha. He will continue sitting quietly on his mound till the dog gets near, when he retreats quietly into his burrow. The dog never can resist the sight of a vizcacha, and never learns that it is impossible to catch one, so this game goes on for ever, to the great enjoyment of the vizcacha.

The birds and beasts and insects who profit by the work of the vizcachas and make their homes on the mounds are on the friendliest terms with them, and, indeed, the foxes take a base advantage of the friendliness of the little creatures. They come into the vizcacha dwellings, and stay there till a 'wing' is given up to them. The good-natured and easy-going vizcachas, however, do not resent this, and may even be seen taking the air with their guests on summer evenings. All goes well till the young ones are able to leave their cells, when the foxes throw off their masks and seize them for dinner, even fighting the old ones first.

For a long while, the vizcachas, being of no use to mankind, were let quite alone, but of late they have grown so very numerous, and the land has been thrown so much more into cultivation, that it has become necessary to destroy them. Efforts have been made to stop up some of their burrows, but their friends learn in some wonderful way of the danger of their comrades, and will come even from great distances to dig them out. Their wonderful powers of endurance enable them to stand a long siege, and they can live without food for as long as fourteen days. If they are allowed to die a natural death in their own inner chamber, the dead vizcacha is left to lie in state, for a few days, after which he is carried out by his relatives, and placed solemnly on the side of the mound.

GUANACOS: LIVING AND DYING

WHEN the Spaniards under Pizarro conquered the great country of Peru, about the year 1520, they found much value set upon the race of *Llamas*, of which four kinds existed in Peru, all of them highly prized for one thing or another.

The llama itself, which is the largest of the four, is chiefly used as a beast of burden, though it can only carry about a hundred pounds weight at a time, and is able to travel no more than sixteen miles a day, or about as much as an ordinary soldier's march. If an extra pound is put on, the llama simply declines to move; and if its driver tries to give it a blow, he will receive something very unpleasant in his face, as visitors to Zoological Gardens know very well! A hundred pounds does not seem a great load for such a large beast, but there are many qualities about a llama which cause him to be employed, rather than many another stronger animal.

First he is *there*, and in great numbers, so that he is to be had for the asking. Then he is easily managed; never wants water for weeks or months together, and lives on any poor kind of grass (especially a sort called ychu) that he can pick up on the sides of the rocky Andes or Cordilleras. His wool is so thick and clinging, that it is very seldom necessary to tie on the load, which sticks on of itself; a pointed claw enables the llama to walk safely over slippery places, even over ice, much better than any shoes would do, and finally, if no other food is to be had, his flesh is quite tolerable.

The llama varies in colour, but is generally of a sort of white. His neck and legs are long. Fond of company, when his pride is not touched he is easily led, and it was no uncommon thing for the Spaniards, when they first entered the country, to meet whole

SPANIARDS MEETING A CARAVAN OF LLAMAS

caravans of llamas laden with silver ingots from the mines of Potosi, travelling under the charge of a single native. Indeed, it has been reckoned that fully 300,000 llamas were employed in this service.

But though the wool of the llama was sometimes used for rough kinds of cloth, it was not nearly so highly

valued as that of the smaller variety of the breed, called
the vicuña, whose hair was woven into the finest
material, reserved especially for the Peruvian nobles.
The vicuñas are little beasts, with soft feet and excellent
appetites, and when the grass on the higher mountains
withers in the summer heat, they come down in search of
the pasture on the moist plains. In every herd there are
generally fifteen or sixteen females to one male, but he
is very careful of his charges, and when they are on
the march always brings up the rear. The little ones
are strong and swift, even from the moment of their
birth; but when the males are quite grown up the
mothers all join together to expel them from the flock,
and the young creatures then form a club of their own,
from which, in their turn, the females are excluded.

The laws of hunting in Peru were very strict, and the
peasants were strictly forbidden to break them. Once a
year the Government arranged a chase on a large scale,
which lasted a whole week, and was shared in by all
the men of the district; but great care was taken that
the hunt should only be held in the same place every
fourth year. Each man had his appointed place and
brought with him a pole and spear, and a weapon called a
bolas, made of two balls joined by a string. This was
whirled round the head and let fly at the animal, and so
skilful were the Peruvians in its use, that the creature
was generally killed at the first blow.

As the hunt went on, the circle of men was drawn
closer and closer, till at the end, nothing was left alive
but the valuable vicuñas and their cousins the guanacos,
who were always held sacred. Then a great shearing
took place—sometimes as many as forty thousand of
these llamas remained to be sheared—and the wool was
stored in the royal magazine till the different kinds
could be sorted and separated. When this was done,
the finer sorts were reserved for the nobles, and the
rest given to the common people, who had a right to

all the flesh of the dead animals. The skins fell to the priests.

The guanaco, which is smaller than the llama, and larger than the vicuña, wanders over the whole of South America, and is to be met with on peaks of the Andes more than twenty thousand feet high, as well as in the bare lands of southern Patagonia, where he is most numerous. He is about four feet high from the shoulder, and seven or eight feet in length, and his wool is of a pale reddish colour and very thick. It is shortest and reddest on the top, and is exactly suited to the cold bare places where the guanaco loves to roam. Like the llamas, they are generally to be found several together, but they are very cautious, and never attempt to eat the smallest meal without placing a scout to give warning of the approach of an enemy. If one is seen—or smelt —for the scent of the guanaco is extraordinarily keen, the scout utters a peculiar penetrating cry, something like the bell of a deer, and the flock instantly make off to a place of safety. But if the enemy to be dreaded is a puma, his scent is sharper and his feet swifter than those of a guanaco, and many are the bodies found lying on the pampas, with dislocated necks.

Like many shy people, guanacos are very curious, and, as has happened before now, their curiosity often ends in their undoing. Sometimes a band will unite to explore some special district, and when they have discovered what they are looking for, they will wheel round as cleverly as a troop of soldiers, and return whence they came in a straight line. They are all good swimmers, and of very accommodating habits; if no grass is to be had, they can go without for a surprising length of time, and if fresh water cannot be got, they content themselves with salt. They are lively and excitable, and may be seen giving vent to their feelings just as human beings do, by making strange noises, and jumping about.

Guanacos are very rarely seen by themselves, but may be met with in flocks varying from five hundred down to six. They are easily tamed, but, unlike most other animals, become more ready to defend themselves in their tame than in their wild state. They will even learn to attack man, and to strike out in a peculiar way with both knees from behind.

One strange fact has been discovered about the guanacos which is not as yet known of any other creatures. When, by some curious and unexplained instinct, they feel that they have received their death wound, or been stricken with their last illness, they leave their fellows, and make straight for one of their dying places, perhaps hundreds of miles away. Some of these dying places have been seen by travellers, in South Patagonia, where they are most frequent, usually near rivers, in the midst of low trees, and thick scrub. Why the stricken beast should take the long and often difficult journey, instead of creeping away like other creatures into the nearest hole or thicket to die, we do not know. It may be an inherited longing for a spot which was originally a place of shelter, or it may be that they are pushed by invisible hands to the grassy refuge that is whitened by their father's bones. What becomes of all the dead animals? Does anybody know? Of the sparrows, the monkeys, the hares? A 'dead donkey' is such a rare sight that it has turned into a proverb. But what about the rest?

IN THE AMERICAN DESERT

CHILDREN who are lucky enough to have read Captain Mayne Reid's charming tales, will remember all sorts of exciting stories of animals; but as so many of the old books have fallen out of fashion, we will give a couple of adventures taken from the 'Desert Home,' and no child who has once looked at these will be content without getting hold of the original volume.

The 'Home' was in the great American desert, which spreads over a large tract of country in the west of Texas and the east of New Mexico, nearly to the foot of the Rocky Mountains. People who live in such places must expect strange sights and sounds, and the battle of the snakes, which the settlers one day witnessed, was the sort of thing one might see at any moment.

Robert Rolfe, his wife and family had gone out West, to find a place where they might build a cabin, and live a happy and peaceable life. All sorts of things had to be thought of before a suitable spot was fixed on, but at last a log cabin was put up, and everyone began to hope that, by the time the winter came, the house would be snug and comfortable.

One day they all started for one of the salt springs which are to be found in such numbers through this region, and took a huge kettle with them, to boil down the salt before they carried it home. They had just finished dinner, and were sitting over the fire, when they heard a blue-jay screaming from a tree near by, and from the tone of her voice they knew quite well that an enemy must be at hand.

At first nothing was to be seen that would explain the bird's alarm, but on glancing from the trees to the ground, they saw a thin yellow body moving noiselessly through the grass. Every now and then it stopped, raised its head, and touched the dry leaves with its tongue, and in so doing it stretched itself out, showing its full length, which was over six feet. At the end of its tail was a loose row of horny substances, which made a horrid sound when shaken, and gave the creature its name of 'rattle-snake.'

Now, of course, no snake in the world could catch a bird if the bird chose to fly away, and the rattle-snake least of all, as it cannot climb trees. But snakes, as everybody knows, have a deadly power of fascination, and the people who were looking on were anxious to see whether the blue-jay would be able to resist the charm, or whether she would fall a victim to his spell.

By this time the snake had reached the foot of a big magnolia, and after sniffing all round the tree, coiled itself up in a great yellow heap, close to the stem, paying no heed to the foolish blue-jay, who had done her best to bring about her own death by the silly noise she was making. However, seeing at last that the snake was paying no attention to her, but only getting ready for a nap, the bird plucked up courage, and flew away to its nest.

A moment after, the rattle-snake made a slight movement, which proved he was not asleep after all. What was he waiting for? A squirrel most likely, for squirrels are the dinner which rattle-snakes like the best. Yes, sure enough, high up in the tree there was a hole, and along the grass was a tiny trail leading straight to the magnolia, and from certain marks on the bark, it was quite plain that the squirrels came and went that way. Now it was close to this trail that the snake had taken up his station.

They all sat with their eyes fixed on the hole, out of which a little head came peeping. It did not see the snake, but it did see the settlers, and did not seem to like the look of them, for there it was, and there it clearly meant to stay. Suddenly the dead leaves of the wood began to rustle violently, and out dashed another squirrel at its topmost speed, making for its home in the tree. Twenty feet behind a long yellow pine-weasel was in full chase.

The squirrel could think of nothing but the enemy behind, and never heeded any possible danger in front, yet, if it had only looked that way, it would have seen something more dreadful even than a pine-weasel. The rattle-snake had suddenly swelled to twice its natural size ; his mouth was opened so wide that the lower jaw touched his throat, and his poisoned fangs were bare. As the squirrel flashed past him up the stem, the snake appeared to move his head slightly, but so little that it did not seem even to have touched the squirrel. Yet somehow, before the squirrel had reached the first branch, it began to climb more slowly, and in another moment stopped altogether. It swayed from side to side as if it had been seized with giddiness, then its claws gave way, and it fell dead into the jaws of the serpent.

The weasel, who in its headlong chase had very nearly rushed upon the same fate, stopped at a little distance, hissing and growling, and evidently half inclined to fight the snake, but at length it decided that this would be very unwise, so, with a final snarl, it marched off into the woods.

When the last hair of the weasel's tail had vanished round a tree, the snake uncoiled himself, and licked the body of the squirrel well over, before swallowing it head foremost.

He was still engaged in this operation when a great creeper with scarlet flowers, hanging about twenty feet above the head of the rattle-snake, began to move in a curious way, and out of the wreaths of leaves and blossoms came a big, black body, as large as a man's arm—a boa-constrictor. It glided down the creeper towards the trunk of the magnolia, taking the greatest care to do nothing which could rouse the attention of the rattle-snake, who, indeed, was wholly occupied in making ready the squirrel to swallow. He had just taken the head and shoulders into his mouth, when the boa-constrictor appeared dangling for a moment by a single loop of his tail; then he dropped, and before the lookers-on had time to see what had happened, both snakes were locked together in a death struggle.

As to size they were very well matched, but the boa-constrictor was thinner, and far more active. It wound and unwound itself round the rattle-snake's body, pressing it close in its crushing embrace; and the rattle-snake was powerless to sting, as it could not get rid of the squirrel. Curious to say, they never fought face to face, but the head of the constrictor had seized the bony rattles of its foe, and with its strong tail was really beating him to death. It was quite plain who was going to win; the snake had no weapons now his poisoned fangs were useless, and soon his struggles grew feebler in the grasp of his enemy, and he stretched himself out, as dead as the squirrel.

In places that are the homes of wild animals, not a day passes without adventures of this kind. One day Mr. Rolfe and his son Frank went down the valley to collect some moss which hung in strips from the branches of the 'live-oak,' and made soft and comfortable stuffing for mattresses. They soon discovered what they wanted, and were very busy about their work, when some black and yellow orioles began making a terrific hubbub in a grove of pawpaws close by. Leaving the moss the two men crept behind a tree, to see what had caused the disturbance. It was some minutes before they found out, then they saw an extraordinary bundle of heads and legs and tails coming slowly along towards the grove of pawpaws. What in the world could it be? Suddenly the great body seemed to divide up into a host of little ones, and behold

WATCHING THE COMBAT

there was an old grey, woolly opossum, the size of a big cat, and her thirteen little white rats of babies! Opossums are ugly creatures, and this one was no better than the rest. Her nose was long and sharp, her legs short and fat, and her tail, which was nearly the length of her whole body, was quite naked. Underneath, she had a pouch like a kangaroo.

When she had got rid of her thirteen children, the old opossum stood still and stared straight up into a tall pawpaw, where the birds were fluttering and screaming more wildly than ever; every now and then making a dive down to the opossum, who took no notice of their proceedings. The two men followed her eyes, and saw an oriole's nest, hanging like a pocket from the top twigs of the tree.

The old opossum saw it too, and uttered a sharp cry which brought all the little ones running helter-skelter from their game in the dead leaves. Some tucked themselves safely into their mother's pouch, two used her tail as a rope, and lay comfortably down in her hair, while others held on by her neck. When seven or eight had gone to their places, the whole mass began to climb the pawpaw. At the first branch the heavily laden animal stopped, and then, holding the kittens one by one in her mouth, she passed their tails twice round the branch, and there they hung heads downwards, looking very funny indeed. When she had disposed of those she had with her, she went back to the ground for the rest, till the whole thirteen were suspended from the branch!

This business done she could now go up the tree with an easy mind, and very cautiously she made her way towards the nest, the birds growing more and more excited the higher she climbed, till their wings almost touched her nose. When she reached the branch on which the nest hung, she stopped doubtfully. It was very thin, and creaked beneath her weight as she moved along it. Clearly it was too dangerous, so she backed carefully till she was safe on the trunk, where she paused to consider what she could do next. All at once the branch of an oak, that stretched exactly out over the nest, caught her eye, and turning, she ran swiftly down the stem of the pawpaw, and up the oak. In another moment she was creeping out on the branch.

When she was right over the nest she curled round her tail, and let herself go. But it was no use. In vain she swung herself backwards and forwards, stretching herself out to her greatest length: the nest was still too

THE MOCCASON SNAKE FASCINATES THE ORIOLES

far away. The eggs she coveted were only a few inches off, but they were as much beyond her reach as if they had been miles away. At last, with a snort of disgust, she swung herself back again, and came down the oak.

The young ones were soon unhooked from their station on the pawpaw, and tucked away as before. Then, evidently in a very bad temper, she took her whole cargo off into the wood.

The birds now changed their note, and after singing a short song of victory, became quite still. Suddenly the fluttering and chattering began afresh, and through the grass came gliding a huge moccason snake. If the birds had only known, they and their nest were safe enough, for the moccason cannot climb trees, but it has other ways of getting at its prey.

The nearer the snake came the greater grew the noise of the orioles, though every circle that they made brought them lower and lower, and closer to the snake. The moccason watching steadily, saw that the spell of his fascination had almost worked, for the birds sometimes actually touched the ground in their flutterings, while their wings moved more and more slowly. At length one stood quite still with his mouth open; but instead of seizing his prey, the moccason suddenly uncoiled himself and took flight the way he had come, while the birds, who had so narrowly escaped death, flew into the tree.

The reason of the snake's strange conduct was the sudden appearance of a peccary or wild hog on the outskirts of the wood, a creature about as large as a wolf, with bristles in place of hair, and sharp tusks sticking out of its mouth. It was closely followed by two young ones who, instead of being dark grey, were a kind of red.

The peccary had not seen the snake, and was not thinking about it, till suddenly she stepped by accident across its trail. The smell of the moccason was quite unmistakable, and she ran about with her nose on the ground, sniffing the scent. At first she made one or two false starts, for, of course, the snake had left a double track; but having settled on the right one, she started off at full speed.

Meanwhile the snake was hastening as quickly as its

natural slowness allowed, towards the shelter of the cliffs, taking care to keep itself hidden as it went in the long grass. But the peccary, coming galloping along with her nose on the ground, almost tumbled over it before she was aware, and both parties drew back and prepared for battle. For a minute or two they eyed each other; the peccary drew back and then came on with a sudden rush, ending with a spring high into the air, which brought her straight on the moccason's back. It was a most curious form of attack, for no sooner had the peccary alighted on the back of the snake, with all its four paws pressed closely together, than it bounded off again. This was repeated two or three times, and then she sprang right on the head of the moccason, breaking its neck on the ground by the pressure of its claws. Once more the thicket sounded with the cry of victory, and the peccary, calling to her young ones, who had taken no part in the battle, ran up to the snake, which she skinned very neatly with her tusks and teeth, before eating the flesh for supper.

But she had not very long to enjoy herself with her family, before she was disturbed. Through the weeds and jungle which grew up to a short distance of the bare spot where the combat between the peccary and the moccason had taken place, came stealing softly a beast with a long thin red body, and a head like a cat. It was the fierce and tree-climbing cougar.

The peccary went on with her supper, quite unconscious that she was being watched by her deadliest enemy, who was calculating his chances of making a successful spring upon her back, for he knew too well what a peccary's tusks were like to wish for an encounter with *them*. Apparently he decided that the leap was too great for his powers, so he turned stealthily back and ran up a tree which cast its shade over the group of peccaries. Then, gathering himself together, he uttered a battle cry, and leapt straight on her neck.

For some time the fight raged—silently on the part of the cougar, noisily on that of the peccaries, for even the little ones ran round and tried to think they were lending a helping hand ; but the peccary had no chance from the beginning, and before long was lying dead on her side, with the cougar lapping her blood.

But strange noises were now heard coming nearer and nearer through the brushwood. The cougar rose quickly to his feet, and tossing the dead peccary over his shoulder, he made off in the opposite direction to that from which the sounds came.

It was too late ; for at the same moment a herd of twenty or thirty peccaries, who had been summoned by the cries of the dying one, rushed across the open, and cut off the retreat of the cougar. In an instant he was surrounded, and flinging down the body of the peccary, he sprang upon the nearest living one, felling it with a blow from one of his paws ; at the same moment he himself was seized from behind and pulled down.

It did not seem possible that one animal, however fierce, could keep so many foes at bay ; but two or three times he shook them off and sprang into the air, only to be caught and dragged back by some watchful peccary. At last, he gathered up all his strength, and with a desperate leap cleared the circle, and made straight for the tree where the two men were sitting, and before they could even cock their rifles, he was crouching on a branch above them, and glaring at them with his fierce eyes.

If ever anybody might be said to be 'between the devil and the deep sea,' it was Rolfe and Frank at this moment ; for if the cougar was above them, below were the peccaries puffing and snorting, and tearing at the bark of the tree.

However, in a few seconds Rolfe collected his senses, and came to the conclusion that the enemy above was more to be dreaded than the enemy below. For peccaries cannot climb trees, but cougars can and do quite easily !

In fact, had it not been for the presence of the peccaries, he would never have waited so long.

They had only one gun between them, and even Frank's bow and arrows were not to be counted on, as he had carelessly left them lying at the foot of the tree, and the peccaries had long since made them into chips. Rolfe therefore told the boy to change places, and get behind him, so that the first brunt of the cougar's attack might fall upon himself. This was done quietly, but with some difficulty, for it is not very easy to pass another person on the branch of a tree.

When they were settled in their places, Rolfe fired at the cougar's head, for the rest of the body was covered by the thick moss. For a moment the smoke prevented his seeing if the shot had taken effect, and he felt as if every instant he might feel the creature's claws in his throat. While it was still too thick for him to make out anything, he heard something falling heavily through the leaves; then a thud and a scream and a rush, and in a minute or two the peccaries trotted away.[1]

[1] These anecdotes are not to be taken as historically true.

THE STORY OF JACKO II.

THE winter of 183— was unusually severe in Paris, in spite of all the predictions to the contrary of Matthew Lansberg, the weather prophet.

Counting on the mild season he foretold, many people laid in but a moderate supply of fuel, and amongst them was the artist Tony Johannot. Whether this was the result of faith in the prophet, or of some other reason into which it might be indiscreet to inquire, the fact was that towards the middle of January this distinguished painter, on going to fetch a log from his wood cupboard, discovered that if he continued to keep up fires in both studio and bedroom his store would barely hold out another fortnight.

Now there had been skating on the canal for a week past, the river itself was frozen, and Monsieur Arago announced from the Observatory that the frost would certainly increase. And the past being a guarantee for the future, the public began to think that M. Arago was probably right, and that for once Matthew Lansberg was mistaken.

Tony returned from his wood cupboard much troubled by the result of his calculations. It seemed a choice of freezing by day or freezing by night! However, on thinking the matter well over as he worked away at his big picture of the hanging of Admiral Coligny at Montfaucon, it struck him that the simplest plan would be to move his bed into the studio.

THE STORY OF JACKO II.

As for his monkey, Jacko II.,[1] a bear's skin folded in four would do famously for him.

The move was effected that same evening, and Tony fell asleep in a pleasantly warm atmosphere, delighted with his happy idea.

On waking next morning he felt puzzled as to where he was for a few moments, but soon recognising the studio, his eyes turned by instinct towards his easel.

Jacko II. was seated on the back of a chair, just at the height and within reach of the picture. For a moment Tony imagined that the intelligent creature, who had lived so long amongst pictures, had at length become a connoisseur, and that, as he seemed to inspect the canvas very closely, he was lost in admiration of the beauty of its finish and details. But he soon found out his mistake. Jacko adored white lead, and as the picture of Coligny was nearly finished, and Tony had put in all his high lights with this pigment, Jacko was busy passing his tongue over every spot where he could find it.

Tony sprang from his bed, and Jacko from his chair, but it was too late. Every part of the canvas on which there had been the smallest touch of white lead was licked bare, and the Admiral himself had been, one might almost say, swallowed whole!

Tony began by flying into a great rage with Jacko, but, on second thoughts, reflecting that it was very much his own fault for not tying the monkey up, he went in search of a chain and a staple.

He fixed the staple firmly into the wall, riveted one end of the chain to it, and having thus prepared for the coming night, he fell to work on his Coligny, and succeeded in pretty well re-hanging him by five o'clock.

Then, feeling he had done a good day's work, he went out for a walk, dined at a restaurant, went to see a play, and got home soon after eleven.

[1] To distinguish him from Jacko I., Décamps' monkey.

On entering the studio Tony was pleased to find all in good order and Jacko peacefully asleep on his cushions. He went to bed and was soon fast asleep too.

Not long after midnight he was roused by such a rattling of old irons that anyone might have thought that all the ghosts in Mrs. Radcliffe's novels were dragging their chains about the room. Tony did not much believe in ghosts, but fearing some one might be breaking in to steal his wood he stretched out his hand towards an antique halberd which hung on the wall. But in an instant or two he discovered the cause of all this noise, and shouted to Jacko to lie down and be quiet.

Jacko obeyed, and Tony made all haste to fall asleep again. At the end of half an hour he was once more aroused by smothered groans and cries. As the house stood in an out-of-the-way part of the town Tony thought some one was being murdered under his very windows. He jumped out of bed, seized a pair of pistols, and ran to open the window. The night was still, the street quiet, not a sound disturbed the peace of the neighbourhood; so he closed the window and realised that the groans came from inside the room. Now, as he and Jacko were its only occupants, and as he certainly had not uttered a sound himself, he went straight to Jacko, who, not knowing what to do, had amused himself running round and round the leg of the table till his chain shortened, and as he continued turning round he found himself suddenly pulled up short by the collar. It never occurred to him to run round the other way, so he only choked more and more with each attempt to free himself. Hence the groans which had disturbed his master.

Tony promptly unwound the chain from the leg of the table, and Jacko, happy to be able to breathe once more, retired humbly and quietly to bed. Tony also lay down hoping for a good sleep at last; but he reckoned without Jacko, who had been disturbed in his regular habits. He had slept his usual eight hours early in the evening and was now quite wide awake. The result was that, at the end of twenty minutes, Tony bounded out of bed once more; but this time it was neither halberd nor pistol which he took in hand, but a whip.

Jacko saw him coming, and tried to hide in a corner, but it was too late, and Tony administered a well-deserved castigation. This effectually quieted the culprit for the rest of the night; but now Tony found it impossible to go to sleep again, so he got up, lit his lamp, and as he could not paint by its light, sat down to work at one of the wood engravings which made him the king of illustrators of his day.

He felt much puzzled all the morning as to the best way of combining peace at night with economy in fuel, and he was still turning the matter over in his mind when a pretty cat called 'Michette' walked into the studio.

Jacko was very fond of Michette because she did whatever he wished, and Michette on her side was devoted to Jacko. Tony, remembering their mutual attachment, determined to make the most of it. This cat, with her thick winter coat of fur, would be as good as any stove.

So he picked her up, and putting her into Jacko's hutch, pushed him in after her, shut down the grating, and went back to the studio to watch through a little hole how things went on.

At first the prisoners tried hard, each after its own fashion, to get out. Jacko leapt against each of the three walls, and then fell to shaking the bars of the grating, regardless of the fact that his efforts were quite in vain.

As for Michette, she lay where she had been placed, and looked all round without moving more than her head; then going to the bars she rubbed first one side and then the other against them, rounding her back and arching her tail, and mewing loudly. Then she tried to push her head between the bars, but, finding all of no avail, she made herself a nest in one corner of the hutch, and curled

K 2

herself snugly up, looking like an ermine muff seen from one end.

Jacko, on the other hand, kept on jumping and scolding away for another quarter of an hour, then finding all his efforts to be useless he retired to a corner opposite Michette's. Being well warmed by all the exercise he had taken, he stayed quiet for a time, but he soon began to feel the cold and to shiver all over. It was then that his eyes fell once more on his friend, so comfortably rolled up in all her warm fur, and his selfish instinct at once prompted the use he could make of her. Quietly he drew near Michette, lay down near her, slipped one arm under her, and passed the other through the opening made by the natural muff which she formed. He then twisted his tail round his neighbour's, and she obligingly drew them both up between her legs. when he seemed quite re-assured as to his future.

Tony, satisfied with what he had seen through the hole, sent for his housekeeper and desired her to prepare food for Michette every day, besides the carrots, nuts, and potatoes always served up to Jacko.

The housekeeper duly obeyed orders, and all would have gone well with Michette and Jacko had it not been for the monkey's greediness. From the very first day he noticed that a new dish was served with his two regular meals, one at nine in the morning and the other at five in the afternoon. As for Michette, she at once recognised her accustomed milk pudding in the morning, and meat patty in the evening, and she proceeded to eat each in turn with that dainty deliberation common to all well-bred cats. At first Jacko left her alone; but one morning, when Michette had left a little of her pudding on the plate, he came up behind her, tasted it, and found it so nice that he quickly cleared the dish. At dinner-time he discovered that the mess of meat was even more palatable, and when he rolled himself comfortably round Michette for the night, he spent some time wondering why he, the son of the house, should only have nuts, carrots, and other raw vegetables, which set his teeth on edge, provided for him, whilst this comparative stranger was offered such tempting delicacies. He came to the conclusion that his master was most unjust, and that he must do his best to restore things to the proper order by eating the pies himself and leaving the nuts, &c. to Michette.

So, next morning, when the two breakfasts were brought, as Michette, purring cheerfully, approached her saucer, Jacko picked her up under one arm, where he held her firmly, with her head turned away from the food as long as there was any left on the dish; then, having had an excellent meal, he left Michette at liberty to breakfast in her turn on the vegetables.

Michette turned over and smelt them each in turn, but, displeased with the result of her inspection, she came back mewing sadly, and lay down by the greedy monkey.

At dinner the same manœuvre took place, but this time Jacko was still more pleased with his idea, for the meat pie struck him as even better than the milk pudding. Thanks to these nourishing meals, and the warmth of Michette's fur, he spent an excellent night, snoring away lustily, and quite regardless of poor Michette's complaints.

Things went on like this for three days, to the great joy of Jacko and the equally great distress of Michette, who, by the fourth day, was so weak that she lay still in her corner without moving. Jacko made an excellent meal, and felt much ill-used when he returned to roll himself round Michette to find his warm muff so much cooler than usual.

The night was colder than ever, and next morning Jacko's tail was frozen hard, and Michette lay at the point of death.

Luckily, on that day, Tony, who had felt anxious on account of the extreme cold, went to inspect his two prisoners as soon as he woke. He was only just in time, for both seemed almost equally petrified, so he took Jacko into the studio, and handed Michette over to the cook, who thought for some time that she was quite dead; but the warmth of the kitchen and judicious feeding gradually restored her, and in a day or two she was herself once more; but nothing would ever induce her to go near Jacko again.

Jacko himself was rather stiff, but he soon recovered his circulation and wonted activity, except in his tail, which remained frozen, and which, having frozen whilst curled round Michette's tail, retained a corkscrew form— a shape unknown amongst monkeys, and which had the funniest appearance you can imagine.

Three days later a thaw set in, and the thaw caused a strange thing to happen.

One day Jacko was perched on the top of a tall ladder in the studio, when a lad suddenly came in bringing back a large lion's skin which Tony had sent to be mounted. The boy had hung the skin over his back, and it partly covered his head; and his appearance, and the smell of the skin, so terrified Jacko, that he turned quite faint and fell down from the ladder.

He was promptly picked up and soon restored to his senses, but in the sudden fall his frozen tail had snapped right off, and Jacko had to pass the remainder of his life a tail-less monkey.

'PRINCESS'

SHE was not actually a king's daughter, as far as I know, but she was so evidently high bred, and had such a superior, aristocratic air about her, that the name seemed perfectly appropriate.

There could be no question as to her high descent and pure blood. It was apparent in every one of her graceful movements, in the exquisite softness and delicacy of her grey coat, the thickness and fluffiness of the ruff she wore round her neck, and the size and bushiness of her superb tail. In a word Princess was a pure-bred Persian cat, and her happy owners, Mrs. and Miss H., took great pride in her possession, and much pleasure in her society.

Indeed, they declared that her understanding was quite beyond that of ordinary animals, and that she quite understood much of their conversation.

One day Miss H. went out to make some calls, and on her return sat down to tell her mother all about her visits. Princess jumped into her lap, and curled herself up cosily, as if to listen to her adventures.

Presently, Miss H. said: 'You have no idea, mother, what a magnificent cat Mrs. Taylor has. It is immensely big, and has one of the most splendid tails I ever saw.'

In a moment, Princess rose, sprang from Miss H.'s lap, and walked to the door, demanding to be let out. It was clearly not for her to stay and hear one of her own mistresses praising the charms of a horrid rival.

Mrs. and Miss H. made acquaintance with a lady whom we will call Miss Gray, and to Miss Gray

Princess took a curiously strong fancy at first sight. If Miss Gray happened to be calling at the house, and Princess chanced to see her parasol or umbrella in the hall, she would hurry off with every sign of delight in search of her dear friend. If several people were in the room, Miss H. would sometimes say, 'Where is Miss

'PRINCESS' AND THE INVALID

Gray, Princess?' and the cat would turn her head towards the lady and go up at once to rub against her.

'Do you love me very much, Princess?' asked Miss Gray, once. Princess replied by looking up affectionately at her, and uttering a most tender 'miau.'

One sad day, Mrs. H. fell ill, and had to take to her bed. She grew worse, and her poor daughter was very unhappy indeed about her. Princess appeared quite to understand, and to enter into all the trouble and anxiety, and went about sad and drooping. The doctor was very anxious that his patient, who was extremely weak, should take plenty of nourishing food; but nothing seemed to tempt her fancy.

One thing after another was tried—soup, jelly, game—all of no use. The invalid declared she could touch none of them, and poor Miss H. felt in despair.

One morning, as she was sitting by her mother's bedside, and trying to coax her to eat something, the door, which was slightly ajar, was pushed open, and Princess ran in quite gaily. She jumped on the bed, and, with an important air, laid down on her mistress's coverlet a bird she had caught and brought her.

Both Mrs. and Miss H. declared afterwards that they were sure Princess thought she had found the very thing with which to tempt a sick appetite.

THE LION AND THE SAINT

IF you should have the opportunity of seeing any large picture gallery abroad, or our own National Gallery in London, you will be very likely to come across some picture by one or other 'old master' representing an old man, with a long beard, sometimes reading or writing in a study, sometimes kneeling in a bare desert-place; but wherever he may be, or whatever he may be doing, there is almost always a lion with him.

The old man with the beard is St. Jerome, who lived fifteen hundred years ago, and I want now to tell you why a lion generally appears in any picture of him.

At one time of his life, St. Jerome lived in a monastery he had founded at Bethlehem. One day he and some of his monks were sitting to enjoy the cool of the evening at the gate of the monastery when a big lion suddenly appeared walking up to them. The monks were horribly frightened, and scampered off as fast as they could to take refuge indoors; but St. Jerome had noticed that as the lion walked he limped as though in pain, and the Saint, who always tried to help those in trouble, waited to see what he could do for the poor animal.

The lion came near, and when he was quite close he held up one paw and looked plaintively at the men.

St. Jerome fearlessly took the paw on his lap, and, on examining it, found a large thorn, which he pulled out, binding up the injured limb. The wound was rather a bad one, but St. Jerome kept the lion with him and nursed him carefully till he was quite well again.

ST. JEROME DRAWS OUT THE THORN

The lion was so grateful, and became so much attached to his kind doctor, that he would not leave him, but stayed on in the monastery.

Now, in this house no one, from the highest to the lowest, man or beast, was allowed to lead an idle life. It was not easy to find employment for a lion; but at length a daily task was found for him.

This was to guard and watch over the ass, who each day carried in the firewood which was cut and gathered in the forest. The lion and ass became great friends, and no doubt the ass felt much comfort in having such a powerful protector.

But it happened, on one very hot summer's day, that whilst the ass was at pasture the lion fell asleep. Some merchants were passing that way and seeing the ass grazing quietly, and apparently alone, they stole her and carried her off with them.

In due time the lion awoke; but when he looked for the ass she was not to be seen. In vain he roamed about, seeking everywhere; he could not find her; and when evening came he had to return to the monastery alone, and with his head and tail drooping to show how ashamed he felt.

As he could not speak to explain matters, St. Jerome feared that he had not been able to resist the temptation to eat raw flesh once more, and that he had devoured the poor ass. He therefore ordered that the lion should perform the daily task of his missing companion, and carry the firewood instead of her.

The lion meekly submitted, and allowed the load of faggots to be tied on his back, and carried them safely home. As soon as he was unloaded he would run about for some time, still hoping to find the ass.

One day, as he was hunting about in this fashion, he saw a caravan coming along with a string of camels. The camels, as was usual in some places, were led by an ass, and to the lion's joy he recognised his lost friend.

He instantly fell on the caravan, and, without hurting any of the camels, succeeded in frightening them all so completely that he had no difficulty in driving them into the monastery where St. Jerome met them.

THE LION RESCUES THE ASS FROM THE CARAVAN

The merchants, much alarmed, confessed their theft, and St. Jerome forgave them, and was very kind to them; but the ass, of course, returned to her former owners. And the lion was much petted and praised for his goodness and cleverness, and lived with St. Jerome till the end of his life.

THE FURTHER ADVENTURES OF 'TOM,' A BEAR IN PARIS[1]

Part I

Décamps and his brother Alexandre were entertaining a number of their artistic and literary friends one evening in the well-known studio, on the fifth floor of No. 109 Rue du Faubourg St.-Denis, in Paris. Thierry had just finished reading a scientific paper on the peculiarities of frogs, of the same species as Mademoiselle Camargo, when the door opened, and the master of a neighbouring café entered, bearing a large tray covered with cups, saucers, teapot, &c., and followed by two of his waiters who carried a huge hamper, in which were a loaf, some buns, a salad, and an enormous number of little cakes of every possible size, shape, and flavour.

The loaf was for Tom, the bear; the buns for Jacko, the monkey; the salad for the tortoise, Gazelle, and the tea and cakes for the guests.

The beasts were very properly served first, and the guests were then told to help themselves.

A few moments of confusion followed, during which each made himself comfortable after his own fashion. Tom carried off his loaf to his hutch, growling as he went; Jacko fled behind some busts to munch his buns, and Gazelle slowly dragged the salad to be enjoyed

---

[1] For the various allusions to artists, authors, and animals in this story, the reader is referred to the *Blue Animal Story Book*.

peacefully under a table, whilst each visitor provided himself with a cup of tea and such cakes as he fancied.

At the end of twenty minutes the teapot was empty and the cakes had vanished. The bell was rung, and answered by the master of the café. 'More!' cried Décamps, and the master of the café bowed himself out backwards, and hastened to obey orders.

Whilst waiting for 'more,' Janin read to the assembled company that interesting account of the early days of Jacko I., with which all readers of the 'Blue Animal Story Book' are doubtless familiar.

The applause which followed this history was suddenly interrupted by a piercing shriek on the staircase. Every one rushed out to see what was the matter, and found the porter's little girl half-fainting in the arms of Tom, who, startled by this sudden interruption, hurried off downstairs.

A moment later another scream, even more shrill than the first, was heard. An old lady, who had lived on the third floor for the last thirty-five years, had come as far as the landing to discern what all the noise was about, and, finding herself face to face with the fugitive, fainted instantly away.

Tom turned back, hurried up fifteen steps, found an open door and burst into the midst of a wedding feast. Here was a hullabaloo! The whole party rose to their feet, and, headed by the newly married couple, rushed to the stairs. In a moment the inhabitants of the house from cellar to attic were standing out on the various landings, all talking at once, and, as generally happens in such cases, no one listening.

At last the story was traced back to its beginning. The girl who gave the first alarm said that, as she was bringing up the cream she felt some one seize her round the waist. The staircase was dark, and thinking she had to do with some impertinent lodger she promptly dealt him a smart box on the ears. Tom had replied by a growl, which

TOM FRIGHTENS THE LITTLE GIRL

revealed his identity, and the girl, horrified to find herself in a bear's paws, uttered the scream which had given the first note of warning. As already said, the sudden appearance of Décamps and his guests had frightened Tom, and Tom's fright had resulted in the fainting fit of the old lady and the rout of the wedding party.

Alexandre Décamps, who was a special friend of Tom's, undertook to make his excuses to society in general, and as a proof of his docility, to fetch Tom wherever he might be, and bring him to make his own apologies. He went into the dining-room, and there he found Tom walking about the table with great dexterity, and in the act of finishing his third tipsy-cake.

Unluckily this proved a climax. The bridegroom shared Tom's tastes, and he appealed for sympathy to all lovers of tipsy-cake. Loud murmurs arose, which not even the docile air of Tom as he followed Alexandre Décamps could subdue.

At the door Alexandre met the landlord, to whom the old lady had just given notice to leave. The bridegroom declared that nothing would induce him to stay in the house another hour unless justice were done him, the other lodgers chimed in as chorus.

The landlord grew pale, as he foresaw himself left in an empty house, and, turning to Décamps, told him that, much as he wished to retain him as a lodger, it would be impossible to do so unless he lost no time in getting rid of an animal which had caused such a disturbance at such an hour in a respectable house.

Décamps, who, truth to say, was getting rather weary of Tom's various scrapes, only hesitated long enough to save appearances. He promised that Tom should leave the premises the very next day, and, completely to reassure the lodgers, he at once took his bear down to the yard, where he made him get into a large dog-kennel. He then turned the opening to the wall, and heaped big stones on the top.

The promise, and the immediate removal of Tom, appeared to satisfy the complainants. The porter's little girl dried her eyes, the old lady paused in the middle of her third attack of hysterics, and the bridegroom nobly declared his willingness to content himself with some other delicacy for want of a tipsy-cake.

All retired to their own apartments, and an hour later everything was as still as usual.

As for Tom, he first tried, like Enceladus, to get rid of the mountain weighing on him; but finding he could not succeed, he made a hole in the wall and passed through it into the garden of the adjoining house.

Part II

The tenant of the ground-floor of No. 107 was not a little surprised next morning at seeing a bear walking about amidst his flower beds. He had just opened the glass door leading to the garden steps with a view to enjoying the same exercise, but he quickly shut it again, and proceeded to examine the strange intruder through its panes.

Unluckily the hole Tom had made in the wall was hidden by some shrubs, so there appeared to be no clue as to where he came from. The ground-floor tenant then remembered having read lately in his newspaper an account of a most remarkable shower of toads which had fallen at Valenciennes, accompanied by thunder and lightning. The toads, moreover, fell in such quantities that the streets and roofs of the houses were covered with them.

The ground-floor tenant raised his eyes, and seeing a sky as black as ink overhead, and a bear, for which he could in no way account, in his garden, he began to fear that the Valenciennes phenomenon was about to be repeated on a larger scale, and that, in fact, Tom was but the first drop of a heavy shower of bears.

One seemed no more marvellous than the other, the hail was bigger and more dangerous, that was all. Full of this idea he looked at his barometer, which stood at 'Rain. Very stormy.' At that moment a clap of thunder was heard, and a vivid flash of lightning lit up the room.

The ground-floor tenant felt that not a moment must be lost. More bears might fall, and he must protect himself against all emergencies. So he at once despatched his valet for the Commissioner of Police, and his cook for a corporal and nine men, so as to place himself under the protection of both the civil and military authorities.

The passers-by, seeing the valet and cook run off in haste, began to gather round the hall door, and to suggest all sorts of improbable reasons for this excitement. They questioned the hall porter, but he knew no more than they did. The only apparent fact was that the alarm came from that part of the house which lay between the courtyard and the garden.

At this moment the ground-floor tenant appeared at the front door, pale, trembling, and calling for help. Tom had spied him through the glass doors, and, accustomed to the society of men, had trotted up to make acquaintance. But the ground-floor tenant, mistaking his intentions, looked on these friendly overtures as a declaration of war, and hurriedly beat a retreat. As he reached the front door he heard the panes of the garden door crack.

Retreat became flight, and he appeared, as I have already said, before the idle crowd, with every sign of distress, and calling for help at the top of his voice.

As usual on such occasions, the crowd, instead of coming to the rescue, dispersed hurriedly, but a municipal guard who was standing near held firm, and approaching the ground-floor tenant asked how he could help him.

The poor man had neither voice nor words under control, but pointed, speechless, to the door he had just opened and the steps he had come down.

The municipal guard understood that the danger lay on that side, and bravely drawing his sword ran up the steps, through the door, and into the ground-floor apartments.

The first thing which met his eye on entering the drawing-room was the good-humoured face of Tom, who, standing on his hind legs, had pushed his head and front paws through one of the panes and was inspecting these unknown regions with some curiosity.

The municipal guard paused—uncertain, brave though he was, whether to advance or retreat; but no sooner did Tom catch sight of him than with a kind of smothered roar he hastily drew back his head and forepaws, and made all possible haste to take refuge in the furthest corner of the garden.

The fact was that Tom had never forgotten the beating given him by the municipal guards on the occasion of that memorable visit of his to the masked ball at the Odéon Theatre. He connected the sight of their uniform with the treatment he had received at their hands, and this being the case, it is not surprising that as soon as he saw one of his enemies appear in the ground-floor drawing-room he made haste to quit the premises.

Nothing is so inspiriting as to see your enemy in flight. Besides, as already said, the guard was not wanting in courage; so he set off after Tom, who, after two or three unsuccessful attempts to climb the wall, had placed himself in an angle, rose up on his hind legs and prepared to defend himself in accordance with the lessons in boxing given him by his friend Fan.

The guard, on his side, put himself into position, and lost no time in attacking Tom according to every rule of art.

After a few rounds Tom dealt his opponent such a blow on the arm that his wrist was dislocated, and the gallant guard found himself at the mercy of the bear.

Luckily for him the Commissioner of Police arrived at

this moment, and seeing this act of open rebellion ordered the corporal and his nine men to come down into the garden, whilst he himself stood on the top step to give orders.

Tom, interested in watching all these ceremonies, let his antagonist escape, and remained standing upright and immovable against the wall. Then began the inquiry. Tom was accused of introducing himself forcibly and by night into an inhabited house, and, further, of having attempted to murder a public functionary. Not being able to produce any witness to the contrary he was condemned to death, and the corporal was desired to proceed to immediate execution, and ordered his men to load their guns.

Then, amidst a profound silence throughout the crowd which had followed the soldiers into the garden, the corporal's voice alone was heard. He made his men go through the full number of evolutions, but when he came to the word 'present' he turned and looked towards the Commissioner of Police.

A murmur of pity ran through the crowd, but the Commissioner had been disturbed in the middle of his breakfast, and was inexorable. He stretched out his hand.

'Fi ——' began the corporal; but before the word was out of his mouth or the bullets out of the guns, a man hastily rushed through the crowd with a paper in his hand.

It was Alexandre Décamps with an order from Monsieur Cuvier for Tom's admission to the Zoological Gardens, under the special care of one of the most eminent keepers.

It came only just in time; but Tom was safe, and Alexandre led him off, amidst enthusiastic applause, to spend his remaining years in dignified retirement.

RECOLLECTIONS OF A LION TAMER

Among my very earliest recollections is that of running and playing, along with other little urchins, in front of a heavy caravan, at whose horses' heads walked my father. We were about to halt for the night, at Laval, which we could see perched on the hill-side in front of us.

The weather was fine, the sun shone brightly, and we ran gaily to and fro, like so many puppy-dogs let out to play, shouting and laughing about nothing at all, as delighted to arrive at a strange place for to-night as we should be to set off to-morrow morning for a fresh one.

Suddenly there arose a cry—a cry of anguish—that still echoes in my ears, mingled with a horrible sound as of crunching bones. Swiftly I turned round; in the place where my father had been stood a group of men, some stopping the horses, some kneeling round a formless mass under the wheels. Terrified and weeping, I ran back as fast as my little legs would carry me, to find that this dead weight was all that remained of my father. In a jolt of the waggon the shafts had struck and knocked him down: one wheel had gone over his feet, the other had crushed in his head; life was extinct. We were fatherless, and my mother was left in sole charge, not only of her little children, but also of the menagerie.

My father had been, first, a travelling pedlar; then, after his marriage, he had started a panorama of scenes from Napoleon's wars, and when the public grew tired of that he obtained some curious animals, and by degrees acquired enough to stock a travelling wild beast show. Fortune seemed to smile upon him at last, and this venture was in a fair way to become a success, when his life was cut short in this sudden fashion.

My mother strove her utmost, for the sake of her little ones, to carry on the business; but what can a woman do alone at the head of a menagerie? how can she cope with coarse, rough grooms, and foul-tongued stable-men? She soon married again, a painter from Nantes, and for two years we lived very happily together, till he, too, died suddenly, and we were left a second time fatherless.

Thinking she was doing the best for us, she took a third husband, an Italian, named Faïmali; hot-tempered and fretful, he was a real household tyrant. For some years I had the good fortune to escape from his ill-humours, being brought up by an uncle near Mayence, and educated by the monks; but, my education finished, I was obliged to return to my parents and travel about with the show, and then my miseries began. The ill-treatment I received at the hands of my step-father! the blows, the cuffs he bestowed on me! The chief cause of his displeasure was jealousy; having been early accustomed to go freely in and out among the animals I had lost all fear. I had served my apprenticeship amongst them, beginning with the wolf and ending with the lion, and passing through all the progressive stages—each less amiable than the preceding—of jackal, leopard, hyena and panther. I had learnt to look upon them all as friends, and where another would see a menace in quivering lips, curling over jaws bristling with white and shining teeth, I saw nothing but a smile of welcome.

Being then fearless, young, slight and active, I interested and attracted the public more than my step-father did, hence his ill-humour with me. When I appeared on the scene rounds of applause greeted me, to be repeated when I withdrew, the sounds following even into the dressing-room. All the applause my step-father ever bestowed on me was a box on the ear!

It was his way, and I suppose he meant well; still, as I did not appreciate that sort of playfulness, I decided that we must part. Useless to ask his permission, it would not have been granted, for if I *was* his rival, I was at the same time the attraction of the show. I drew the public, and thus increased the receipts.

Therefore it must be done silently, secretly. One fine day, accordingly, after an unusually severe beating, I slipped away to seek my fortune for myself, with the sum of $2\frac{1}{2}d$. in my pocket. It was not a large sum with which to begin the world, but I was fifteen, strong, and full of confidence in myself and in my good star.

The first day I spent wandering about on the hill-side, enjoying my liberty and the fresh air, and subsisting on a loaf of bread and a draught of water from a spring. Next day I spent in similar fashion; but at the end of it even this frugal fare had exhausted my slender resources. After that I wandered about the country, getting a few scraps at one farm, a drink of milk at another, and sleeping at night in the stables of a third. At the end of a week, however, I had enough of this vagabond existence, and went down to the nearest town in search of work.

In the market-place a gaping crowd of rustics surrounded a strange sort of vehicle, whose owner, a quack dentist and vendor of miraculous ointment, was holding forth in praise of his wares. I immediately proffered my services in any capacity whatsoever, and was promptly engaged as head-groom and drum-major, being, needless to say, the sole and only occupant of both posts. I filled these functions with satisfaction to myself during about six months, for though badly paid, I was well fed and independent. One evening, however, my master informed me suddenly that he had no longer sufficient means to carry on the business, and therefore had no further need of my services. He sold his wretched screws of horses, burst my big drum, and I was again turned adrift on the world, with no larger fortune than before. This time, however, I had gained experience and credit; by means of the latter I bought a pedlar's pack, and, like my father before me, started on my travels.

These, however, did not last long, a woman who kept a grocer's shop insisted on adopting me and taking me into her business; but I quickly wearied of that uncongenial occupation, and, thanks to my good looks, and to my capacities as orator—acquired while with my late master, the dentist—I was soon engaged as showman in a travelling waxwork exhibition. My duties there were light and easy, but the inactive life, in the midst of these inanimate figures, was wearisome and monotonous to one of my stirring nature.

The only break in the sameness of my existence was on Sundays, when a young and charming girl named Maria, an orphan, spent the day with my master and his family. As the weeks rolled on into months, and these in turn succeeded each other, little by little we fell in love, and when at last circumstances separated us we found how indispensable we had become to each other. What caused us to part was the arrival in the neighbourhood of Bernalco's menagerie. It was a fine one, and the animals it contained were not only numerous but formidable of aspect. All my old passion for wild beasts instantly revived, and it was, at all events, an active, stirring life, such as my nature required, I offered my services, and these being promptly accepted, I thought I had at last attained my desires. But no, I was not yet satisfied. I soon found the animals to be so tamed, so subdued, so docile, that there was no excitement, no risk in going amongst them. They aroused themselves from their sleep and went through their exercises so submissively, so mechanically, that at times I had a mad desire to seize them by the mane, and to cry out: 'Try to be fierce, can't you?'

And so we went on, slowly travelling towards the south—each day each performance exactly like the preceding; till at last a red-letter day dawned in my career. It happened at Bayonne, in this way: The performance was just about to begin, one afternoon, the band was tuning up, and the spectators were already assembling in crowds at the foot of the wooden steps leading to the arena, when the cry arose: 'Ather has escaped!'

Now, Ather was a young royal tiger, noted, perhaps with some slight exaggeration, for his ferocity; the only one, in fact, of all our animals not a sluggard. Everyone had seen him prowling to and fro in his cage, rubbing himself against the bars, at times lashing his tail in a fury, while his bloodshot eyes darted flame and fire. In the menagerie he was tractable enough, but at liberty, out of doors—his prey all ready to hand—who could answer for the consequences?

In one instant the public fled helter-skelter into the houses, up on the roofs, some even climbing the nearest trees. As for me, feeling that my long looked for opportunity had now arrived, I straightway set off on his track, under the burning afternoon sun. I had been a considerable time in his pursuit, when a window was cautiously opened, and a voice said, almost in a whisper: 'He is there;' while a hand pointed to the half-open door of a locksmith's workshop, which, in contrast to the brilliant sunshine outside, seemed a cavern of darkness. In I plunged—at first I saw nothing; but after a few seconds, becoming accustomed to the darkness, I perceived the fugitive, with flaming eye and slavering jaws, crouching in a corner all ready to spring. Another instant and he would leap on me, seize, and rend me. I forestalled him, however, and it was I who leapt upon *him*! What a combat ensued! what roaring, raging, foaming, scratching! Fortunately it was of short duration, or it would have been all over with me.

Seizing him with my large, strong hands by the scruff

of the neck, I slung him over my shoulder, and, neither staggering nor stumbling under this enormous burden, I bore him in triumph back to the menagerie. You may

I SEIZED HIM BY THE SCRUFF OF THE NECK

imagine that I was pleased with myself, and will probably suppose that my fortune was made and my position assured for the rest of my life. Anything but that; my master was on his way into Spain, and, as I did not know Spanish, he had no further need for my services. Here was I turned adrift again, and not much further advanced than when I received the same treatment at the hands of my late master the quack dentist. At Bordeaux I gained a livelihood for a while by making and selling little balloons. Then a naturalist took me with him to Havre, where I found another menagerie, Planet's, in which I succeeded in obtaining a berth. My new employers treated me well, and I threw myself ardently, passionately, into the study of my work, observing, reflecting, and laying the foundations of my future career. Now that I felt my prospects becoming more assured I ventured to let my thoughts return to Maria, who, I felt, had become indispensable to me. A fortunate chance bringing us with our respective shows to the same place, I dared to tell my secret to her, and to ask her to unite her lot to mine. Her consent I readily obtained; but her adopted parents withheld theirs till I should have gained a position of my own. I determined to conquer every obstacle in order to win her. With this end in view, I thought to introduce a new element into the profession; to become, in fact, not an exhibitor of tamed and subdued animals, but a real tamer of the fierce and unvanquished. By one means and another, therefore, I contrived to become the possessor of a cargo of freshly caught forest-bred animals, whose ferocity had hitherto daunted the courage of the most audacious. I dared them, I defied them, and I quelled them. It was a real battle, that might at any moment end in slaughter. When I entered the cages it was often doubtful if I would ever come out alive; and frequently I have heard sighs of relief from the spectators when I emerged safe and sound. The expenses were heavy, if it were for the keep of the animals alone; and often I have gone without dinner myself in order that they might fare well; it was better so to do, for if I had put them on short commons they might have avenged themselves by devouring *me*!

At last my success was generally acknowledged. I paid off all my debts, and became absolutely my own master. Not till then did I dare to demand again the hand of Maria, which this time was granted me. We were soon married, and then we proceeded very humbly to start a menagerie of our own. We began at Lyons with a monkey named Simon—whose antics were a constant source of amusement—some serpents and some crocodiles, also two or three boas, which last we kept in our bed-room in the little hotel where we lodged ; but as they continually slept the sleep of the just they disturbed no one. From time to time we received additions to our establishment. One evening there arrived a cargo of crocodiles, which, until they could be properly caged, were deposited, still in the cases in which they had travelled, in a kind of cellar opening on the courtyard. There Maria and I, accompanied by men with lighted lanterns, went to work to unpack them.

All creatures of the crocodile tribe are totally wanting in grace and charm, and cannot safely be recommended as household pets. Unlike all other creatures, they have the lower jaw immovable, while the upper one closes on its prey with a spring, both jaws being furnished with no less than 175 teeth. They are clothed in an invulnerable coat of mail, and their tail is a powerful weapon, that shatters, mangles, destroys, everything that comes in contact with it. Added to all these other attractions they are at no time of amiable disposition, particularly after ill-treatment or in confinement, and if they escape they become the most ferocious of creatures. Now ours had just undergone long imprisonment on board ship, and one of them escaped.

What a scrimmage then took place ! The men made for the door, all the lights went out, Maria and I climbed on a table, two of its legs gave way and we were hurled on to the floor beneath, vainly groping along the walls for the door, and pursued by terrible growlings and flappings of tails against the floor and furniture. At last I found the door handle, and we were safe. But safety alone was not enough. I had paid more than 150*l.* for the brutes, and I could not afford to let one escape, nor to let them destroy each other. Taking a lighted torch in my hand I returned to the fray, and presently succeeded in imprisoning under a seat the monster, who measured no less than four yards long.

Shortly after this unpleasant incident we lost one of these costly pets by death; injured at the time of its capture, it suddenly fell ill and died at the end of twelve hours. Not only was this a great loss to us—for it was one of our finest, and had cost, alone, 80*l.*—but we were due next day at Seyne, near Toulon, with all our beasts, and how could we appear without the advertised number? A happy idea struck me: I went to a naturalist, whom I knew in the town, and asked him to come to my aid by stuffing the animal, and thus passing him off as a sleeping crocodile. What a night we spent—cutting him up, cleaning him out, stuffing him, and putting him together again; but before the morning dawned our task was accomplished, and stretched on a little stage the creature had all the appearance of a *bona-fide* sleeping crocodile.

Taking one of the liveliest of the small ones, I made him furiously flap his tail and open wide his terrible jaws, purposely exaggerating his ferocity, and at the same time giving the usual explanations out of the natural history books. When I saw that I had sufficiently excited my audience I turned towards the stuffed crocodile, and said with trembling voice:

'Oh, what you have just seen is nothing! Now if this sleeping crocodile would but awake, *then* you would see something *really* terrifying. One blow of his tail, and this shed would be shattered to atoms; as to his jaws, he has only to open them—— But you shall see for yourselves; you have only to say the word and I will awake him at once.'

'Oh, no!' exclaimed the audience with one voice, 'by no means awake him!' and this was just as lucky for me. By means of this same ruse I succeeded, day after day, in drawing fresh spectators. At Marseilles alone, where I made a stay of three weeks, after paying off all expenses I had a clear gain of 100*l.* This sum I laid out in fresh purchases—a lioness named Saïda, two hyenas and two

THE LION TAMER OFFERS TO WAKE THE (STUFFED) CROCODILE!

wolves, next a panther, and, later, a bison and a black bear.

These last were the cause of a disagreeable scene that took place one night at Avignon. After an unusually heavy day, I was sleeping peacefully in my waggon, when I was aroused by an appalling noise, bellowings of pain and furious growlings, accompanied by terrible blows, which made the walls and floors resound. Evidently a fierce fight was going on somewhere. Hastily dressing, I hurried to the scene, and discovered that the black bear had contrived to overthrow the bars which separated his cage from that of his neighbour, the bison, upon whom he had fallen, and, hugging him bear fashion, was now, with his long sharp tusks, pitilessly devouring his hump, buffalo hump being esteemed a delicacy. The danger was imminent, for if in their struggles the door should become open there was no end to the consequences that might ensue. Throwing myself between the combatants I held one by the neck while I sent the other flying back to his den. Thus the peril was averted, and next day it was as if nothing had happened, except that the bison was humpless.

The success of my menagerie had now become so generally acknowledged that, after visiting all the principal towns in the south of France, I crossed the frontier and went south into Italy, where each stopping-place was the scene of fresh triumphs. In Florence the king and all his court were present at a performance, where I surpassed myself in daring and audacity. The king applauded louder than anyone, and afterwards desired that I should be presented to him in order that he might congratulate me himself. Encouraged by my successes, I determined to push on to Rome. There a terrible catastrophe came near taking place. I was seated, one afternoon, at the desk taking the tickets, just as the performance was about to begin, and the enclosure was already crammed with people, when suddenly there were heard heart-rending cries, succeeded by furious roarings, and frantic shrieks of 'Help! help!' In an instant I was in the enclosure, where I found general panic, women fainting, men yelling, and all eyes turned in the direction of the lions' cage, where Venturelli, one of my men, hung suspended in mid-air from the claws of four lions; one was devouring his arm, blood from which spurted in all directions.

To raise the bars and slip into the cage was the act of a second. How I was not torn in pieces myself I know not, for I was defenceless, with neither firearms, stick, whip, nor weapon of any kind, but my two powerful fists; hitting out right and left with these I ordered the lions to their dens. They obeyed me, and slunk away submissively, letting fall their hapless victim, who was picked up almost lifeless and conveyed to the hospital, where, however, he recovered from his wounds. I asked him afterwards how he came to let himself be caught.

'Ah, sir,' he answered, 'am I not your pupil? As I was passing near these gentlemen' (for he always spoke very respectfully of the lions) 'I thought I would like to pat them; three were sleeping, but the fourth awoke his comrades, and if you had not been there, sir, I should surely have been made mincemeat of.'

It was at Rochefort that I received my first wound: a lion in a sulky fit defied me, growling and showing his gleaming tusks. I lashed at him with my whip, and he sprang upon me. I darted aside, but not in time to avoid a blow from his heavy paw, the claws of which tore open my thigh. I punished him; but he was, perhaps, to be excused, for the performance that evening took place under peculiar circumstances; there being no gas we were obliged to light with candles, and no doubt this unusual illumination irritated and annoyed him, for no one can imagine how small a thing will put out a wild beast.

Lyons was the scene of a terrible disaster. While there I received from Africa a superb lion, recently captured and still untamed, packed in a solid cage, and that enclosed in a special van, on which was a label with a full description of its formidable contents; no risk need have been run by anyone coming in contact with it. But, unfortunately, while the train which bore the monster to its destination was being shunted in a siding, a cattle drover, named Picart, was foolhardy enough, in spite of all warnings and precautions, to risk his life, first in offering a piece of bread through the bars, and next, as his sleeping majesty took no notice of this affront, in attempting to pat the lion on the head. Then arose the king of beasts in his wrath, and, quick as lightning, the unhappy man's arm was seized upon, crunched and snapped off, with no more ado than a dog would make over a chicken bone. His piercing cries promptly brought the railway officials to the rescue, armed with pitchforks and iron bars; but, alas! it was already too late, the monster was licking his lips over the sanguinary morsel, while the unhappy victim of his own folly was writhing in agonies, which mercifully soon ended in death. I gave a performance next evening for the benefit of the widow and children, during which I entered the den of the rebellious lion and publicly chastised him.

After several weeks' successful performances at Lyons I proposed to make a stay at Marseilles; and, in order to convey my living freight thither, chartered a train of forty trucks. Decidedly Lyons station brought me bad luck: it now witnessed a tussle with an elephant, who positively declined to enter the car destined for him; no persuasions, no coaxings would induce him to budge— the more I pommelled him the more he resisted; finally we were obliged to resort to force. A strong rope was passed around his legs, at the other end of which was a gang of nearly a hundred men, who, dragging and tugging, only succeeded in embarking the unwilling passenger as the train gave the warning whistle for departure.

The white bear and the elephant were the cause of the first really serious danger I ever ran. In the midst of an exercise the white bear, irritated by some trifle, suddenly threw itself on the elephant, who, surprised by this unexpected and unmerited attack, set to work slowly and methodically to repel his assailant. Feeling myself the natural peacemaker, I threw myself between the combatants, with the result that I became the assailed party. Hugged in the close embrace of the common enemy, I should soon have ceased to breathe: a few minutes more and it would have been all over with me; but, summoning all my remaining strength, I hammered with my two strong fists on the brute's nostrils till the blood, flowing in torrents, blinded and bewildered him. Profiting by this state of affairs, I slipped from his grasp, and seizing a stout ash stick that stood handy, I belaboured him soundly, which speedily had the effect of calming him, and soon he was ambling sanctimoniously along as if nothing had happened.

All these accidents only increased my passionate love for my career. In answer to all remonstrances I maintained that having triumphed over so many perils I was bound to continue triumphing. But I reckoned without the little proverb of the pitcher that goes often to the well. One ill-fated day the forebodings of my friends (and also rivals) were fulfilled, and I made my last appearance as a lion tamer.

One hot July afternoon, at Neuilly, I perceived, as soon as I entered the menagerie, a certain excitability among the animals, about which, however, I did not excite or disturb myself, putting it down to some atmospheric cause, and feeling confident that, should any commotion take place, I should be able to quell it. The afternoon performance passed without a hitch; when the evening one began, I entered the cages as usual, and there passed tranquilly before me—each in his turn—the first, second, third, and fourth lions, and next, the two white bears. Finally, I was left alone with Sultan, the same who, a short while before, had devoured the arm of the unhappy cattle drover. He was a fine black African lion, eighteen years of age—the prime of life among his tribe. He could at no time truthfully be accused of good nature, and I perceived at once that evening that he was in one of his worst humours. When I ordered him to leap the bar, as usual, he sulked in a corner and refused. When I cracked my whip at him he growled, and the more I urged him the louder he growled, showing his teeth and beating the air with his heavy tail.

To withdraw from the cage and leave him master of the field would have been to acknowledge myself beaten, and that would not have been in keeping with my character. I determined to conquer him, and, in this determination, I advanced a step forward. Now it happened that I was then suffering from rheumatism, particularly in my knee, and as I stepped a sudden shoot of pain caused the knee to bend, and me to fall to the ground!

Instantly I knew that I was lost. In a moment Sultan was upon me, one heavy paw resting on my head, while with the other he tore, gashed, clawed my quivering flesh.

On all sides resounded cries, shrieks from terrified women, frantic calls for help from men. I, and I alone, uttered no sound, I knew the necessity for calm and the danger of the slightest false move. Seizing the raging animal by the skin of the throat, I twisted it with all my force, in hopes of strangling him. By degrees his frantic movements began to cease, and his powerful muscles to relax. Suddenly he turned his head; something behind was taking his attention off me. What was happening was this: two of my men had succeeded in entering the cage, and, with red-hot irons, were attacking his flanks. Profiting by the momentary respite, I managed to raise myself to a sitting posture, and I shortly found myself upright again; once on my feet I felt that I was safe, and would resume my position as master. Summoning all my resolution, I advanced on the rebel, and peremptorily ordered him to his den. I compelled him, though sullenly, to obey me, and would next have proceeded to chastise him with an unsparing hand, but I yielded to the clamour of the public and allowed my rescuers to lead me from the cage. The only concession I obtained was permission to make my bow before the audience, whose sympathies I had thus involuntarily aroused. I admit that my clothes were hardly in a fit condition for a public appearance—I was covered from head to foot with blood and sand; one sleeve was in rags, the lappets torn off my coat, and the collar altogether missing. Then I had to submit to be taken home, put to bed, and let the doctors examine my wounds. These numbered no less than seventeen : my arms were lacerated, my shoulders were beaten black and blue, while my throat—and this was the most painful of all—was torn open. Three weeks was I obliged to keep my bed, it was even reported that I would never leave it. But, at length, I regained my liberty, and the first use I made of it was to revisit the menagerie. I had grown a thick beard, the collar of my coat was turned up, and the brim of my hat slouched down over my face, rendering me almost unrecognisable ; but I had hardly set foot in the arena than Sultan scented me and greeted me with an angry growl. Flattening himself against the bars, he stood, his claws stretched out, ready to spring upon me ; his eyes blazing with rage, and every bristle on end, while the evil expression on his horribly contorted face, plainly said : 'What! not dead yet ? Wait till the next time I get hold of you !' I could with difficulty be prevented going into his cage and paying off old scores at once ; but I was constrained to content myself with visiting all my other friends, while Sultan followed my every movement with an angry eye.

I was impatient, as soon as my health should be completely re-established, to begin my performances. When that happy evening at length arrived I entered the cages as calmly as of old, and the exercises were gone through without any impediment. When it came to Sultan's turn, to my surprise he contented himself with growling, and did not attempt to attack me. I kept none the less on my guard, for I knew that he was revengeful, and was only awaiting his opportunity, the result proving that I was not mistaken.

This opportunity arrived one evening in October, when, the directors of the Zoological Gardens in Paris being present, I naturally was on my mettle, and wished to let them see of what I was capable. Sultan seemed to grasp the situation, and a repetition of the former scene took place. This time, however, I was prepared, and stood firm on my legs, not flinching under his onslaught. It was a veritable combat, but I was the stronger, and though armed with nothing but a whip, I succeeded in forcing him, outwardly tamed, to obey me.

Needless to say that from that day our relations continued somewhat strained, and at every performance I foresaw a fresh explosion. It required all my strength, all my determination, to cope with his animosity, which each day seemed to augment, and every time that I left him I was worn out in mind and body from the sustained efforts.

At length my hour struck, though not in the way I had anticipated. Suddenly I became aware that my right cheek had lost all feeling, my upper lip was drawn up, and my right eye glazed. The doctors were hastily summoned, and pronounced me to be paralysed! The strain of my continued efforts had been too great, my machinery was worn out, and I, so active, so greedy of life and movement, was condemned to perpetual repose. Farewell in future to all emotions and ovations! farewell to all the pleasure daily received from the applause of the public!

No more should I appear among my animals, to lord it over them, as of old; in future I must be reduced to the rank of simple spectator, condemned to watch others perform feats taught them by me, and enact the *rôle* I had created for myself, and had filled with such signal success throughout so many years.

SHEEP FARMING ON THE BORDER

THE sheep possesses all the virtues. Those whose lot is cast in the hills of the Scottish Border (as is the case of the writer) will know how to pity the poor sheep who have to find their food on the hills from December to the month of May—hills which are sometimes deep in snow, and which at best grow browner and deader day by day, often till June is reached.

Yet the sheep work on, often doing with no other food than they can pick for themselves, or, perhaps, a little hay at the best. This is in what is considered a good winter. But generally, in the course of these months the monotony is broken by storms of snow or wind—or both united—which produce terrible suffering to the poor animals, and to their masters.

My great-grandfather, who lived on the same farm as we do, kept minute diaries of these things, and from these, as well as from the stories of the 'Ettrick Shepherd,' I have got my information.

The first big storm recorded is that known as the 'Thirteen Drifty Days.' It was about 1672, and must have occurred soon after sheep farms were set going on the Border. Now, a steady fall of snow, unless it reaches a great depth, is not a very great misfortune, as the sheep scrape away on the steep hillsides with more or less pluck (if you watch them you see a marvellous difference in degree), and they manage to get enough to live on, as the grass is always fresh beneath the snow.

But if a wind gets up and drift sets in, one sits by

one's fire shuddering to think of the wreaths or drifts piling up outside. Well, in this year of storm there had been a long spell of snow and frost, and the surface of the snow had hardened, and the sheep were weakened by want of food. In the end of January a change seemed coming, and the shepherds were rejoicing to think of relief. Little they thought what the change would be! The wind rose and drift set in, nor did it cease night or day for thirteen days. The accounts doubtless lose nothing in the telling; still it seems certain that in all those days the sheep never broke their fast; nor was the drift constant from one quarter, for the wind shifted so continually that the shepherds knew not how to dispose the poor animals for shelter. On the ninth and tenth days the dead grew so numerous, from hunger and the most intense cold, that the shepherds built up dykes or walls of dead sheep in a half-circle to shelter the living. It availed but little; and on the fourteenth day, when the storm at last abated, nought remained on any farms but these walls of dead sheltering a small flock, all likewise stiff and cold. One happier experience is recorded in our long diaries.

A certain Robbie Scott, of Priesthaugh, in Upper Teviotdale, never left his sheep day nor night all through the weary storm. He scraped away what snow he could where the drift had left ground comparatively bare, and he led the sheep to where the rough tops of heather afforded them some little food. A fine fellow he must have been, and of most wondrous endurance; but, worn out at length, on the thirteenth night, he went away to get the sleep he could no longer do without. By morning it was thawing, so his sufferings were not in vain; and later he was rewarded by his sheep bringing eight score lambs, which was more than the whole district altogether could show.

But the greatest storm on record is that of 1794, known as the Gonyal storm—no one knows why—when the thaw did almost as much damage as the actual storm itself. The ground had been covered with hard frozen snow for some time before it began, and the shepherds were all keeping their weather eyes open in expectation of a blast. The day before the storm was thick, dark, and piercingly cold, but without a breath of wind, so that no one had an idea from which direction the tempest would come. One old man, out of his own experience, said that wherever the first opening appeared through the fog the storm would burst; whereat his fellow-shepherds hooted, for just then a south wind sprang up, and the opening appeared in the north! Nevertheless the old man was right, for, towards midnight, with a roar like thunder, the hurricane broke with a blinding drift from the north. This lasted for about a week; but to give you some idea of the strength of the blast, I must tell you that two hours before daylight it was impossible to get out of any door facing north, so deep were the drifts outside. In a short time the whole aspect of the country was changed; dykes, of course, had vanished, valleys were levelled, burns which, in the morning, had been swollen to the size of rivers, had in many places disappeared, and even trees were buried entirely out of sight.

So you can, perhaps, understand a little the difficulty younger shepherds, who were new to the district, had in rescuing the sheep on this occasion; for they recognised their whereabouts only by landmarks, and were dismayed to find that everything had completely gone. But all the experienced hands had set out before daylight—with their hats tied firmly on their heads, their plaids sewn round them, and a good flask of whisky in their breast pockets—in search of the sheep. They plodded thus, three or four of them together, in single file, each man leading in turn, for the fury of the blast was such that no mortal could stand up against it for more than ten minutes at a stretch. It seemed an almost hopeless errand; on arriving at the place where the sheep should have been there was no sign of any living creature! The collies were then set to work, and it was extraordinary to see how quickly they pounced upon the place where a sheep lay buried; and one old dog, Sparkie, is said to have smelt out several at a depth of no less than fifty feet below the snow! The sheep were all living when found; but those that were very deeply buried felt the sudden change into the bitter atmosphere above, for, after bounding away in delight at their release, they were almost instantly paralysed and fell helplessly upon the snow, where they remained many hours before recovering the use of their limbs.

When the thaw came the rivers rose so suddenly that many of the poor weakened creatures could not get out of the way in time, and there is a curious record of the 'throw up' in the Solway which I quote here: '1840 sheep, 9 black cattle, 3 horses, 2 men, 1 woman, 45 dogs, and 180 hares, besides a number of meaner animals.'

In our own experience things are better: there are more roads, and the railways are of much help in many districts; yet the elements remain as before, and we still have our anxieties. I can call to mind being able to walk over dykes on the snow wreaths, and days of drift when one could not see the course of the Teviot lying just below us. Such was a great storm on Old Year's night in 1874, when six trains were snowed up for two days at the head of Gala Water. Such, again, was a short sharp storm in March 1889. It came on very suddenly, the wind being so violent as to overturn two loaded trucks on the railway near our farm in Liddesdale. The poor sheep just ran before the blast on this farm, going to the head of a deep *cleuch* or glen for shelter; there they could get no further, and were 'smoored,' or buried in the soft snow. We lost 31 on that occasion; but close by, on the Northumbrian border, losses were heavier, for, as ill-luck would have it, the storm took place on the very day the shepherds had all gone to Rothbury for their dog-licences; so the sheep were not gathered, and

DIGGING THE IMPRISONED SHEEP OUT OF THE SNOW

on one farm eleven score were lost! The track of that storm was only about five miles broad.

Again, in 1894, we had a disastrous storm, and all the sheep on our hill farms had to be driven in to our home farm on the Teviot. There they remained for six long weeks, while the ground was caked completely with ice—not a blade of grass or tuft of heather to be seen. They were all, 2,400, fed twice a day by hand on hay. It was curious to see how, when the first breath of 'fresh' came into the air, all the hill sheep stopped eating, and every nose turned in the direction of home with loud and prolonged 'baas;' and I do not know whether shepherds or sheep were most delighted to return to their wilds.

When the joyful day came, it reminded one of the flight out of Egypt to see the long line of sheep and shepherds wending over the hills. Little do our friends, who come to us in summer days, like the swallows, understand how different our winter life is. It has its discomforts, its many anxieties; but it also brings one face to face with nature in a way which does one good. It is grand to force one's way up the hill after a wild storm, and see the snow piled up and blown into all kinds of queer shapes and caves, till one can believe oneself in the Arctic Circle. It is good to see master, and men, and dogs all working together on the quest for buried sheep, feeling about with long poles in likely places till the dogs come to the rescue and scent out the sheep. If the snow is dry and powdery, sheep can live three weeks easily beneath it; but if it is soft it will very soon smother them; in which case great is the anxiety to get them to the light of day.

Such are some of the not uncommon events of Border life—not very remarkable, not very blood-curling; but bringing with them more of hardship than most things in every-day life.

WHEN THE WORLD WAS YOUNG

It is always very difficult for us really to feel that people in other places are working and playing exactly as we are doing ourselves, and that when we are dead everything will go on as if we had never been alive at all. But it is even harder for us to believe that for more thousands of years than anyone can count, the earth went on its way round the sun without numbering one single man among its inhabitants.

Not that our little planet was empty and silent, because men were not there to shout and clamour. Anyone looking down from the moon would have seen our world very much as we see it now. There were mountains and seas, trees, and flowers; there were wet days and fine days, high tides and low ones. To be sure, the observer sitting in the moon would not have been looking at the very same mountains and seas that we gaze at now. At one time, countries, which are now dry land, were covered by an ocean; at another, great tracts, that are at present islands, were joined to the continent itself, while, on the other hand, peninsulas (such as India) were divided by a sea from the mainland. In some cases, mountain ranges had not been formed at all, and the rivers ran in very different courses from what they do to-day.

Well, all these seas and continents were the homes of vast numbers of creatures, some bearing a strong likeness to the animals and reptiles with which we are familiar, others that would be absolutely strange to our eyes.

They did not all live at the same time either. One race would hold sway for more ages than we can guess, and then would die out, perhaps affected by some change of climate, and by-and-by another would take its place, also to disappear when its turn came.

Now, how can we know anything at all about animals which died thousands of years ago? In two ways. From their bones (long since become like a stone in substance), or the impressions of them which have been preserved in the rocks, and from the bodies which have sometimes been found quite complete, with skin, hair, and even eyes, in the frozen marshes of Northern Asia.

But the discovery of creatures in this condition is very rare. In general, scientific men who study the subject have to be satisfied with the skeleton, or with detached parts of the frame, and with this help they have worked wonders. One of the most important things in building up the history of fossil animals is the teeth, and with the aid of these it is possible to find out whether the dead monster fed upon flesh, or upon herbs and leaves, or even if it preferred the wood of the branches. A lightness in the upper part of the body, combined with a small head and short forelegs, tells us that the quadruped could rear itself up on its hind legs, like a kangaroo, while in creatures of the elephant kind, which own a long nose or proboscis, we shall find that the neck is so short that it could not reach its food in the trees or on the ground without help of this sort.

Most of these animals lived long long ago, thousands of years before we have any idea of; but one or two survived till a race of men inhabited the earth, or at any rate some parts of it. The best known of these great creatures is the mammoth, which was very like an elephant in shape, and like him had huge ivory tusks, curving inwards and upwards, instead of being comparatively straight. Sometimes the curve nearly made a complete circle, as in the case of a mammoth skeleton now in

SECURING A MAMMOTH

the St. Petersburg Museum, where the tusks measure nine feet six inches; but a semicircle was more common. The mammoth skeletons are usually over nine feet in height, and fifteen feet in length, and when we add muscles and skin, we shall have a very large beast indeed.

The modern elephant is only to be found in hot countries, and is confined to Africa and to India. The mammoth, on the contrary, preferred a cold or temperate climate, and roamed all over Europe, North America, Siberia, and the northern part of Africa. There is scarcely a single English county, except perhaps Cornwall, where its bones have not been found, in the soft clays and gravels and soil washed down by the rivers in the far-off days, when the earliest race of man appeared on the earth.

How strange it would seem to us now, taking a walk along the wooded banks of the Thames near Oxford, to stumble suddenly on a gigantic mammoth, tearing down the sweet young branches with his trunk! He must have looked a huge monster, indeed, with his powerful tusks, often nearly eleven feet long, and his thick coat adapted to face the snows of England and Russia, and the still greater cold of North Siberia. Over his dark grey skin the soft brown wool curled closely, and, above that, was an outer garment of long, almost black hair. Big and clumsy as an elephant is, a mammoth was bigger and clumsier still; but he was by no means the only great animal that found England in those times a pleasant place to live in; for, in many instances, the bones of the hippopotamus and a woolly rhinoceros are to be seen buried beside him, while lions, tigers, and hyenas had not yet wandered to the south.

In those days, as in these, the elephant tribe, of which the mammoth was one, fed on vegetable substances, and even in Siberia, where such enormous numbers of their frozen remains have been discovered, there was obviously *some* sort of food for them. Birches, willows, and fir trees of various kinds, grew then, as now, in those bleak countries, and when the creatures became tired of eating soft things, they had only to uproot a tree, or tear off one of the branches, and crunch up the wood between their strong teeth. Of course, in African forests, the size of the trees often baffles even the strength of an elephant; but in northern climates, such as Siberia, few could stand against a mammoth, the weight of whose tusks commonly amounted to 320 lbs. !

Now it seems wonderful to us that, after so many ages have passed, we can still find the skeletons of these animals, and, indeed, this can only happen in certain ways. In order to preserve a skeleton or even a whole body, it is absolutely needful that it should be kept shut off from either air or water, or not only its flesh, but its bones, will in time crumble away and vanish. This occurs when the animal dies above ground, or is drowned in some lake or river with a sandy, gravelly bottom; and in rocks made up of these substances we shall find but few fossils, or traces of plant and animal life. But if the bed of the lake should happen to be made of mud or clay, or something into which neither air nor water can penetrate, the body of the creature which has got stuck in swimming, or has been somehow caught fast and held, will gradually sink down till he is entirely covered. By-and-by the mud which wraps him round will have become solid rock, keeping within it one of the secrets of a world gone by. Peat will also preserve bodies that have fallen into it, to be dug out, ages after, fresh and young, and—in the case of men and women—with even their clothes undecayed; but one of the most usual means of preservation consists in freezing the bodies, and thus excluding the air.

The great frozen marshes of the north of Siberia teem with remains of mammoths, which have either died on the spot or been carried down by the floods of the mighty rivers. In warm summers, or during heavy gales, these marshes become thawed or broken up, and sometimes one of the huge creatures that has been lying buried, for anything we know, since the days of the Great Pyramid, or even of the first Emperor of China, may be seen floating on the stream. On one occasion a fat, comfortable mammoth, thirteen feet high, with a thick hairy coat, and wide open eyes, was found standing where the earth had given way under him in one of the marshes in north-east Siberia. He had been frozen in the spot where he fell, and had remained there, no one knows how long, till the whole surface of the land had been torn up by the raging waters of the swollen river. The Russian who discovered the mammoth longed to bring it home; but the body, when exposed to the warm air, soon began to fall away, and all he could do was to cut off the tusks, and examine his food, of which traces were still existing in the stomach. By these he made out that the mammoth had feasted for the last time on young fir cones and pine needles, and then, well fed and happy, had gone to his death. The same fate very nearly befell his discoverers too; for, in their excitement, the men did not notice that the ground was giving way under them also, and had not the boat been luckily at hand, the river Indigirka would have carried men as well as mammoth out to sea.[1]

In searching for remains of fossil animals, we must never forget that the topmost rocks are *always*, except where they have been heaved up by accident, the newest and latest formed, and, from this fact, it is possible to tell which creatures lived at the same time, and which succeeded the other. Now, ages before there were any mammoths on the earth there existed a monster very like him in appearance, but differing from him in three ways. First, he had no hair on his body; then his teeth were simpler than those of the mammoth, and though, like him, the animal lived on branches and trees of various kinds, he could grind rougher and coarser food. Lastly, instead of

[1] *Extinct Monsters.*

one pair of tusks, many of the species had two, one in each jaw.

This variety of proboscis-bearing or long-nosed quadruped of the elephant tribe was called the Mastodon. He lived in America as well as in Europe, Asia, and Africa, and could suit himself well to any climate, though, from the many remains that have been found in the temperate zones, he seems to have disliked extremes, either of hot or cold. The mastodon was a huge creature, the skeleton measuring as much as eleven feet in height, with long straight tusks that have been known to stand out as much as ten feet beyond its head. From the fact that stone arrow-heads have been discovered lying round the skeleton in America, and from stories told by the Indians, it seems likely that the mastodon was living in the New World, at any rate when the earliest men peopled the land; but on our side of the Atlantic it had probably died out long before.

Anyone who examines the skeletons that have been pieced together by those who have made bones their study, will be struck by two things—the immense size and clumsiness of the dwellers both on earth and in the sea, ages and ages before man was dreamed of, and also by certain resemblances with several forms that survive up to the present moment. Besides the animals with long trunks, there are monsters with long heads, all stuck with bony nobs, and in shape like the rhinoceros. These skeletons are mostly found in North America, and teach us that the beasts to whom they belonged must have been very nearly as big as elephants, to whose legs theirs bear some likeness. Their bodies were very heavy and awkward, and their eyes small; they had an odd number of toes on their hoofs, and very small brains. Altogether, it is easy to understand how, when the rain descended, and the floods came, in those far-off times, long before the woolly rhinoceros was feeding with the mammoth on the

MEGATHERIA

banks of Siberian rivers, the stupid, awkward animals should have been unable to place themselves in safety, and got swallowed up in the mud of the lake.

Then, too, but much later in date, the great pampas or plains of South America were the home of the ancestors of the Sloth tribe—Megatheria by name—animals eighteen feet in length, whose bones are found in the river deposits. As in all animals that can, stand on their hind legs, the lower limbs and back were immensely strong, while the thick tail acted as ballast; the very thigh bone is three times as thick as that of an elephant. Like the modern Sloth, the Megatherium had no teeth in front, but it probably possessed a long and flexible tongue, which it used to curl round branches of trees and tear them down. It was also able to dig its sharp, powerful claws into the trunk of a tree, and with a mighty heave of its body to loosen the roots, and by repeating this process three or four times the tree would fall to the ground, and the particular morsel on which the Megatherium had set its heart would be within its reach.

Further back still we find that birds and mammals have not yet come into being, but, instead, their places are taken by a strange kind of flying reptile, whose wings were more like those of a bat than a bird, and often measured twenty-five feet. The name of Pterodactyl has been given to this extraordinary creature, which resembles some of the queer fancies men used to carve on churches rather than anything we ever see now. The pterodactyl had teeth, but no feathers, and could swim as well as fly. As to its food, we guess from its teeth that it lived chiefly upon fish, though it may sometimes have swooped down, when flying, on little animals, or even have pecked at fruit.

But besides the pterodactyls, there existed at the same period, which has been called the Age of the Reptiles, vast swarms of creatures whose forms seemed to be made up of a large number of other species. In many ways they were most like crocodiles, but in other respects, again, they remind us of ostriches. To this class Naturalists have given the name of Dinosaurs, from two Greek words, which mean 'terrible lizards.'

All the tribe were alike in one way, for they had four legs; but in some the structure of the bones shows that the Dinosaur could, when it chose, stand upright, while other varieties, such as the Brontosaur, must have been compelled, or at any rate must have preferred, to walk on all fours. This monstrous beast was about sixty feet long, its skeleton has always been found on the bank of a lake or river, and it probably fed on water plants. It had a long neck, which would enable it to rear its head out of the water and see whether the coast was clear of its enemies, and a long tail, which was a great help in swimming. But when on land it must have been difficult indeed for an animal of such huge bulk to get out of the way when attacked, and still more difficult for it to escape detection, as every one of its tracks measures a whole square yard.

About the same time that the Brontosaurus was wallowing among the reeds of the lakes and rivers which covered the tract of country now called Colorado, one of his distant cousins might have been met with any day in the Weald of Sussex, had there been anyone living on the earth to take a walk! This particular reptile has been given the name of Iguanodon, from a peculiarity of its teeth. The largest kind known is thirty feet long, from its nose to the end of its powerful tail, and when walking, as it always did, on its hind legs, was as tall as a very big elephant. From the hollowness of its limb bones it was able to move more lightly than some of the other animals whose bones were solid throughout, and this was very necessary, as, unlike many of these old lizards, the Iguanodon had no sharp knots or spines on its skin to ward off the attacks of its flesh-eating foes. So, when standing as high as it did, it saw one of these huge monsters in the distance, it had time to get out of the way, either in the water, where its strong tail and hind legs would soon carry it out of reach, or it could find shelter in some hiding-place on land. The Iguanodon itself was quite a harmless creature, with a smooth skin.

STEGOSAURUS

It had hands with four fingers, and a sharp, spiky sort of thumb, whose use has not yet been discovered. The toes on its back feet were only three, but they made up in size and strength what they wanted in number.

A great contrast to the smooth-skinned Iguanodon was the Stegosaurus, traces of whose skeleton have been unearthed at Swindon, but are found far more frequently in the Rocky Mountains. It is, perhaps, quite the most curious in shape of all these strange old animals. Its body forms an arch, with a pair of long solid legs not far from the centre, and another pair of quite little ones near its small head. Right down the middle of its back, stretching from its head to its spiny tail, was a ridge of huge bony plates, like colossal ivy leaves, the centre ones measuring two or three feet across. It seems to have been about twenty-five or thirty feet long; its sense of smell was very acute, its eyes were large, and could absorb much light, and it *ought* to have been very clever, as it had two sets of brains, one in the usual place, and the other, ten times bigger, near the thigh. As may be imagined, the Stegosaurus (or 'lizard with a roof') was very heavy to move, and most likely found it pleasanter to pass most of its time in the water, which, being of more weight than the air, would support its great bones better. But when on land it could defend itself from its enemies by the help of its tail, which had four pairs of strong sharp spikes, calculated to keep the most bloodthirsty animal at bay. Its own food, as shown by its teeth, was soft juicy plants.

There is no time to say much of the largest of all the Dinosaurs, which has been found in America, and measured more than eighty feet. Its thigh bone alone was taller than a man, and if it walked upright it would certainly have been thirty feet high. Nor can we linger over the fish lizards, which came before all these, or the lobster-like creatures that lived before *them*, or over the crocodiles, some eighteen feet long, found in the new red sandstone and later rocks, or over the tapirs, or many more. But we must just glance at a few birds which are now extinct, partly through the merciless hunting down by man, and partly owing to natural causes, with which he has nothing to do.

As far as can be gathered from the rocks, the birds (which did not come into being till the great order of the reptiles had mostly died out) were a good deal less numerous than the creatures who had gone before them. But this may be partly accounted for by the fact that their bodies, being lighter, would more easily float on the surface of lakes and rivers, and would be eaten by fishes, or decomposed by the air, instead of being sealed up in mud, like those of larger and heavier animals.

The *very* earliest kind of bird that has so far been found at all—it was in a Bavarian rock of late limestone, and is known as the Archæopteryx—resembles, in many respects, the family of reptiles. It has, to be sure, a long jointed tail, and teeth in its jaws, and other features in common with them; but then it possesses feathers, even on its tail, and the brain of a bird. Teeth were not at all uncommon in the jaws of these early birds, and the long-billed, fish-eating, Hesperornis, of North America, had a whole set that grew afresh when the old ones fell away. The Hesperornis was between five and six feet high, and is found in the chalk rocks. It was a famous diver, and had wings of a sort; but whatever use they may have been on land, they certainly could have been of none either in air or water.

In New Zealand there existed, until comparatively lately, several 'running' birds, of the kind of which the ostrich, the cassowary and the emu are the last specimens surviving in the world. The most celebrated of these is the Moa, whose bones are found only along the banks of streams at present flowing, showing that the surface of the land has not changed since they were buried there. The Moa could not have been less than fourteen feet, as its leg bone is nearly three times as long as that of a man. Some of the tribe were slight and swift, others like the Dinornis Elephantopus, shorter and stronger. The wing bones of all were so small as to be hardly noticeable, and their bills were invariably short. Whether they could sing or not we do not know—probably not. In some cases a few bones, with feathers attached to them, have been discovered, and from these feathers, combined with a long neck and small head, we gather that the Moa must have resembled an emu or cassowary in appearance.

When New Zealand was first discovered the Maoris found the country, greatly to their surprise, to be nearly empty of land animals belonging to the mammal class, although it swarmed with running birds. Some of them, like the great Dinornis, were as tall as an elephant; but, large or small, their wings were always very tiny and quite useless, and their bones, developed by much running, particularly strong.

The nearer we get to the history of the earth as we know it, the more numerous become the birds, some of which, though now extinct, have lived on till recent years. The remains of a huge bird, called the Epiornis, which in size rivalled the great Dinornis, have been found among the soil brought down by the rivers of Madagascar. Very often a huge egg has lain beside these bones, measuring thirteen or fourteen inches across. This must surely have been the 'roc's egg,' which the Genius refused to give to Aladdin, which was six times as big as that of an ostrich, and capable, says Professor Owen, of containing 148 eggs of a hen!

Travellers in the Indian Ocean during the seventeenth century have left us some interesting tales about a short fat bird, then inhabiting Mauritius and the neighbouring islands, known as the Dodo. And it might have been living there yet, had it not been for men's insane passion for killing. The Dodo was rather bigger than a swan, with a short stumpy tail, decorated, like the little wings, with a bunch of soft feathers like those of an ostrich. Its legs were very short also, and this fact, combined with the weight of its body, rendered it difficult for the Dodo to escape from its pursuers. The flesh seems to have been more appreciated by Dutch sailors than English ones, if we are to judge from the description of our explorer, who declares it 'was better to the eye than to the stomach.' The last Dodo was seen in 1681.

When we visit a zoological gardens, it is impossible not to be struck with the fact that certain of the animals that we are looking at seem strange to our minds, as if they had come from a world of which we know nothing, and belonged to a state of life we could never understand. We may be gazing at all the beasts equally for the first time, but we know exactly what to make of a lion, a puma, or a zebra; while an elephant, a rhinoceros, or a

PTERODACTYL

kangaroo, fills us with thoughts that we can hardly explain even to ourselves.

Now, perhaps these vague feelings arise from these creatures really representing the life of such ages ago that nobody can even venture to guess at a date. As will be seen, from the short account given above, of animals that are now extinct, they none of them were exactly like the beasts and birds to be met with now, but they were like enough to them to show that they belonged to the same race. We all have felt a curious sensation when we have met an old gentleman or lady who persists in wearing the dress of their youth, or in clinging to habits long out of date. Well, this is precisely the impression produced by animals who have modified their ways and appearances as little as possible in conformity to a new state of things. Nature, as we learn if we study her, never works in jumps. She takes into consideration the kind of world the creature has to live in, the kind of food he has to eat, the kind of enemy he has to fight with, and everything about him is fitted for this special life, and this only. If conditions change, he slowly and gradually, but surely, changes with them. Some animals take much longer to adapt themselves than others, just as the Chinese have stuck to their own ways for thousands of years, while, in a quarter of a century, the Japanese have made themselves more European than the dwellers in Europe. Now, the badger, the elephant, and many more, are the Chinese of the kingdom of animals. The very sight of them makes us put our clocks back, and try to fancy what the earth was like in those far-away days. As we have seen, the elephant race lived under various names, in different regions from those where it dwells now, and developed a suitable skin-covering to protect it from the cold. At one time a great beast, in shape like an elephant, but with a certain relationship, too, to the Tapir family, wandered about a large part of Europe, and passed much of its existence in rivers or lakes. The lower jaw of this Dinotherium bent downward, and ended in two heavy tusks, which would only have been an encumbrance on land, but may have been very useful in grubbing up the roots of plants from the bottom of the river. Or he may have dug his tusks firmly into the bank and pulled himself out of the water with their help.

Then, as soon as the rhinoceros quitted the cold regions of the north, and went to live in Africa, he dropped the woolly coat that had protected him, and appeared from henceforward in his dark grey skin, which is much less becoming. As to crocodiles, the oldest known form, found in the new red sandstone, could have looked but little different from our friends of to-day.[1]

It is an established fact that large animals more quickly become extinct than small ones. Their families are fewer, to begin with, and they need more food and water; it is also more difficult for them to hide, and to escape from their enemies. For these reasons, among others, vast hordes of huge monsters have died out, and given place to smaller ones, both in land and sea. And no doubt, if the world goes on long enough, other changes will take place; the old order of things will be swept away, and men will some day be puzzling over the skeletons of cats and the bones of canary birds.

[1] From Owen's *Palæontology*; *Manual of Palæontology*, by Nicholson and Lyddeker, and Hutchinson's *Extinct Monsters*.

BATS AND VAMPIRES

It would be difficult to find any collection of Ghost stories which did not contain one or two tales of Vampires—horrid creatures that steal out of their graves at night to suck the blood of human beings. They make one's flesh creep to read about, but of course they are not alive, and never were.

Now, among the great bat tribe there are most likely several kinds who really do what the stories tell of the Vampires. Indeed, there is one species of big bat, with wings two feet wide, and a horny, prickly tongue, which is known to people who study natural history as the Spectre Vampire. Poor bat, it suffers, as is not uncommon, for the faults of others, for in reality it cares nothing for human blood and has never sucked anybody.

Still, even if we cannot believe all the blood-curdling stories told by travellers in South America and some of the Pacific Islands, as to the proceedings of the Vampire bat, they are very interesting to read, and are true to a great extent about others of the tribe. It is not everybody, fortunately for themselves, that could be sucked by a bat, and no doubt the creatures soon find this out, and fly off to a more promising victim. A curious account is given of their ways by a certain Captain Stedman, who spent five years on the north coast of South America, a long while ago, and he declares that he himself had fallen a prey to their bloodthirsty appetite. According to Captain Stedman, when a bat intends to suck you, he flutters slowly to the ground, and stands by your feet, fanning his wings slowly all the while, to keep you cool and comfortable, and to prevent your waking. Then he

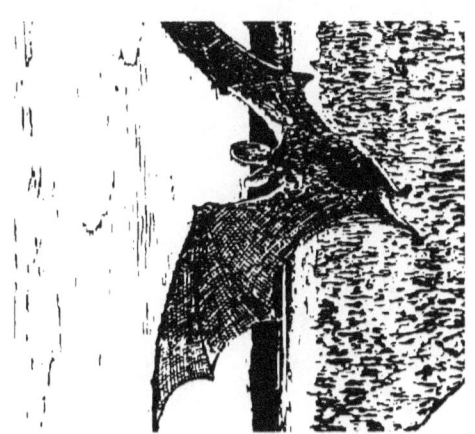

bites a tiny little hole in your toe, not bigger than a pin's head, and from this he sucks till he can suck no more, sometimes after his meal he finds himself too heavy to fly; and sometimes when the morning dawns the sleeping victim is found to be dead.

Cattle, says Captain Stedman, these blood-suckers prefer to attack in the ear, and the best remedy for the wounds is to plaster on the ashes of tobacco.

The common bat which we see darting about in summer evenings, so rapidly that it is difficult to be sure anything has passed at all, goes to sleep all through the winter. In this state it needs no food, but lies in some dark place, hanging head downwards by one of its feet. When the warm weather begins and insects are heard humming round, the bat wakes up too, and flies after them. For though bats will sometimes eat other things, insects are what they like best. Many of them are full of intelligence, and can easily be tamed. They will attach themselves to their masters, rub their heads against them, and even lick their hands. But in general they are not welcome guests inside houses, and are certainly very disturbing to have in one's room at night.

Most bats are of a dark colour, but strange stories are told of their being found of a brilliant scarlet. In each of these cases that have been noted the animal had chosen an odd place for its winter sleep, for it was found inside a tree which was perfectly smooth all round it, and there was nothing whatever to show how the bat came there. One of the trees was a wild cherry, in a wood on the Haining Estate in the county of Selkirk, and was cut down by a woodman, who was felling trees for fences, in the year 1821. The other tree was a pear, cut down near Kelsall five years later, but in both trees the place where the bat was hanging was just large enough to hold him, without much room to spare. Neither bat seemed in the least put out at his rough awakening, but spread its wings and sailed gaily away in search of its breakfast.

THE UGLIEST BEAST IN THE WORLD

MOST people would agree, if they were asked to vote, that the ugliest and clumsiest of all animals is the rhinoceros. Even the hippopotamus shines by comparison, frightful though it is, because, for one reason, it is a water beast, and a water beast can never manage to look so nasty as a land one.

To begin with, the rhinoceros' shape is heavy and awkward, and the horn, right in the middle of its face, does not add to its beauty. Then its eyes, instead of being large and soft, as they so often are in a wild animal, look mean and small, though they are several times the size of those of a man. But, worst of all, its skin is hard and hairless, and looks as if it would come off in scales. Oh, there is no doubt that a rhinoceros is a very ugly beast indeed!

The tribe is divided broadly into two kinds, and is now seldom seen north of the Zambesi river. The white rhinoceros, who must look even more unwholesome than her black fellow, is timid, gentle, and fat, and eats nothing but grass. The black rhinoceros is thin, fierce, and very cautious; but both alike take care never to stray more than seven or eight miles from a river, as they cannot go for months without water, like the eland and some kinds of gazelles.

But whatever we may think of them, even rhinoceroses are not without their friends and admirers; and chief among these are a race of birds, which are never happy unless they are sitting on their broad backs. If by any chance the bird misses its rhinoceros, while the great creature is feeding, as he always does at night, it will call until the clumsy form appears in the first rays of the dawn. The bird also keeps a sharp look-out for any possible danger ahead, for though the rhinoceros's ears are very sharp, his eyes are not, so it is lucky for him that there is somebody at hand who can make up for his deficiencies. In fact, so closely are both bound up together, that when the Bechuanas wish to describe a person they cannot do without, they call him 'my rhinoceros.'

The black rhinoceros is smaller than the white, and, in spite of his heavy body, can run faster than a horse. He is given to sudden fits of passion, nobody knows what for, and then he will burst out into loud snorts, and dash at the nearest bushes with his horn, sometimes digging for hours at the ground round the roots, till he has pulled them up and worked off his bad temper both at once. Perhaps his favourite food, the branches of the tree called the 'wait-a-bit' thorn, which grows to the height of twenty feet, may be irritating. Unlike other animals, the two horns of the rhinoceros do not grow into the skull, but are attached to the skin, one behind the other, and when the animal is dead can easily be taken off with a knife. Rhinoceroses are dirty creatures and love to roll in mud, as their skins constantly show. They stand or lie about in the shade all day long, and in the evening steal out somewhere between nine and twelve to the nearest fountain, and after they have drunk their fill, they go for a long walk. It is very funny to see them taking out their young. The little rhinoceros always walks in front, and if the mother gets the scent of an enemy, they both break into a sharp trot, and the mother guides her child by keeping her horn against its side, and pressing it in the direction she wishes to go. In the case of a white rhinoceros, this horn is about three feet long, but that of its black cousin is much smaller.

Fifty or sixty years ago, rhinoceroses were a great

deal more common in the South of Africa than they are at present, as they have been forced by hunters further and further north. The natives used to chase them with stones and assegais, and so hungry or greedy were they, that even a dinner of the beast's tough flesh was acceptable. Like all animals with hoofs, the rhinoceros feeds on bushes and plants, but this does not prevent his being very fierce when attacked, or from trampling under his great feet anything or anybody that happens to cross his path.

The Namaquas are very cruel in their manner of hunting the rhinoceros, but when once the animal is wounded, and they think it can safely be approached, they try, if possible, to climb on its back, and to thrust a lance into a fatal spot behind the shoulder.

One day a man had just succeeded in getting on to a wounded black rhinoceros above its tail, when the creature started up with a roar, scattering its enemies, who fled for shelter behind a tree. But the tree was not large enough to hide them all, and in a moment the rhinoceros was rushing towards them, tearing up the ground with its horns. The men sped away in all directions, till one of them, getting angry, stopped, and looking the rhinoceros full in the face, called it by an ugly name. The rhinoceros, surprised at this behaviour, stopped too, and stared at the Namaqua, who, gaining courage, became still more abusive. His words seemed to have a power that all the stones had lacked, for the animal turned round and began to beat a retreat, when the Namaqua seizing its tail, sprang on its back, and aimed a deadly blow on its shoulder.

It is wonderful how well the men can aim with their assegais, which are light-throwing spears, with long iron heads. Some bushmen, going on a hunting expedition in their own country, found the fresh trail of two rhinoceroses, and at once set about making their preparations. They first built up a rough stone hut near the place, where one of the men could lie hidden, with two assegais at

hand, while the other went off in search of the animals. After 'becreeping' them, as it is called in that district, for some distance, the bushman saw two clumsy forms lying asleep under a grove of trees, which proved to be a young rhinoceros and its mother. He threw a stone to wake them, and when they jumped up in a rage, he threw a second. The mother looked round to see who the rude person could be that was disturbing her midday nap, and perceiving the bushman, made a dash for him. He had barely time to rush to the nearest tree, and had hardly begun to climb it, when the enraged beast came up, and drove her horns right into the tree and straight between the man's legs, thus giving him time to draw himself higher up out of her reach. She then turned and, followed by the calf, made off toward the stone hut where the other bushman lay hidden. As she passed his assegai touched her shoulder, and after staggering a few steps, she fell dead. The man then aimed his other assegai at the calf, and as it too dropped instantly, he came out of his shelter, while his friend, running up, jumped on the back of the old rhinoceros, and exclaimed, shouting for joy: 'Now I see you are your father's son this day.'

THE GAMES OF ORANG-OUTANGS, AND KEES THE BABOON

THE first Europeans who visited some of the large islands lying close to the Equator—Borneo, Sumatra, and several more—were astonished at finding the woods full of a huge creature which they took for some time to be a man. It was very shy, and disappeared into the thick, dark depths of the forest directly it caught sight of a human being, so it was not very easy to make out exactly what it was like. However, the white men were curious, and also persevering, and at length they were rewarded by seeing one of the largest kind pass by, while they were peeping from behind a bush. No, it certainly was not a man, not even a savage; but how very like one! To begin with, the animal as often as not walked on two legs, and had no tail, while the palms of its hands and the soles of its feet were hairless. The arms were immensely long, and could be used as legs, and the height of a full-grown specimen was sometimes as much as eight feet. This is the animal now known as the Orang-outang.

The whole tribe are wonderfully quick in their motions, and when they are put on board ship can swing themselves about the ropes and rigging in a way that surprises even a skilful sailor. They are affectionate and good-natured, and very intelligent, being able to copy the actions of their masters so closely that, at a little distance, you could not tell which was which. A small orang-outang was brought over to Holland in 1776, but died when it was seven months old, most likely finding the climate too cold for it. Her appetite was very good, and she was seldom known to refuse anything offered to her; but her favourite food was carrots, parsley, and strawberries. Still, she would accept meat, fish and eggs, which she ate very neatly, and was very fond of wine, particularly of Malaga, sometimes drinking a whole bottle at a sitting. During the voyage this clever little lady would make her own bed as well as any housemaid, first shaking up the hay and then getting it all smooth before arranging the bed clothes.

Another of the tribe which was brought from Borneo about forty years later, seems to have been stronger, and to have had a longer life. His captors did not know anything about orang-outangs, and instead of leaving him loose on board the ship, where he would have been perfectly happy, they cooped him up in a cage. However, like other prisoners, he managed by cunning and perseverance, to break through his bars, and ran joyfully up to the top to the mast-head, but by-and-by hunger brought him down, and he was chained up to a strong stake. But one is not a monkey for nothing, and the knot which fastened the chain to the staple was soon undone, and flinging the chain round his shoulder, and taking the end in his mouth, he was off again to his place of refuge.

At last they decided that he had better be left alone, and then there was no end to the games he had with the sailors. None of them could run up the rigging as far as he, or if by good luck or a trick one of them did catch him up, it was nothing for him to fling himself across to a rope hanging thirty feet away; and let the sailors shake the rope as hard as they could his wrists never gave way.

Voyages in those days were very slow, and there was plenty of time to play. Besides, the ships often waited some time at the various ports to take in fresh provisions, and how thankful everybody must have been to get on shore again! The first place that this particular ship put into on its way home was Java, where the orang-outang took up its quarters in a huge tamarind tree. There he at once proceeded to make a comfortable nest for himself by plaiting twigs together, and then twisted in leaves to make it soft. Here he would sit all day long, with his head just peeping out, and if any one passed by with fruit in his hands our friend always went down at once to beg for a bit. At sunset, which comes at six o'clock on the Equator, he punctually went to his own quarters, and at six next morning, when the sun rose, he knocked at the door of his master's hut to ask for breakfast.

Being accustomed to sleep on top of a tree, the moment he was left to himself on board ship he looked about for a place high enough to please him, and of course nothing short of the mast-head would do. Having decided on his bedroom, the next thing was a bedstead and coverings, and for this purpose he got hold of a sail, which he was very careful to spread perfectly smooth, and in this sort of hammock he lay down, drawing the upper part of the sail over his body. Sometimes it happened that all the sails were in use, and then the clever creature would either take the blankets from one of the sailor's hammocks or steal one of their jackets. When the ship got as far south as the Cape of Good Hope the poor thing began to feel very cold, and when he woke in the morning would fling himself shivering into the arms of the sailors, and stay there till he got warm again.

It seems odd to find a monkey drinking tea and coffee, and indeed preferring them to any kind of liquid; but during the voyage, if he could get hold of them, he would take nothing else. This taste, however, died away as soon as he came on shore, for in London he showed a decided liking for beer and milk, though, at a pinch, he would accept wine, or even brandy.

All the long months at sea his master amused himself

ORANG-OUTANGS EATING OYSTERS ON THE SEA-SHORE

by trying to play tricks on the monkey, with a view to discovering how much sense and cleverness he had. To test this his captor would put some fruit in his pocket, and climb up to the mast-head as if to take observations. But anyone who attempts to match himself with a monkey is sure to get the worst of it. As if by instinct, up came the orang-outang, and grasping the ropes with the long toes of one of his feet, he would hold fast his master's legs with the other, and with one of his hands, while the other hand was searching in every pocket. On other occasions he would drop on his master from a height, which must have been very dangerous, or meet him at the bottom, from which there was no escape. Once the man really seemed too clever for the animal, and that was when he tied an orange to the end of a rope, and jerked it up and down, out of the creature's reach. After clutching at it repeatedly without success, the orang-outang pretended complete indifference, and, turning his back, climbed slowly up the rigging. Then he suddenly turned, and springing forward, seized the rope. When this failed, he lost his temper and shrieked with rage, and at length dashing at the man who held the rope, he held his arms tightly, till the coveted treasure was hauled up.

The boatswain was his chief friend on board, and they used to 'mess' together, although neither gratitude nor good manners hindered the guest from sometimes stealing his host's grog and biscuit. After dinner he left the table and sat at the door of his cabin, like a Frenchman on the boulevards enjoying the coffee.

Towards some little monkeys that came on board at Java the orang-outang gave himself great airs—at least as long as any of the sailors were by. Indeed, it was generally thought that he felt a great hatred towards them, especially after he had been one day caught (just in time) in the act of throwing a cage, with three of them in it, overboard. But that was most likely because he could not get hold of some food that had been given them. If he could get the cabin boys to play with he was perfectly happy, but if not (and nobody was looking on) he would put up with the little monkeys, though the games on his side were rather half-hearted. The little fellows, on their part, were much flattered at his notice, and whenever they were let out at once went to wherever their big cousin might be.

In general the orang-outang took all the strange sights and sounds that met him in his new life very coolly, but eight big turtles that were taken in off the Island of Ascension were too much for his courage. As soon as he caught a glimpse of them he tore up to the highest part of the rigging, uttering a squeak of fear, and though at length his curiosity brought him down low enough to catch a peep of them, nothing would persuade him to come quite close. The only other time that he showed any of the same sort of fear was when he saw white men naked (which was quite new to him) bathing in the sea.

Many are the stories of pet monkeys, both orang-outangs and other kinds—putting their masters to shame by sitting over their heads in church, while they were preaching, and imitating every movement, till the congregation was nearly beside itself with laughter. But perhaps no anecdote ever told about the species shows so much intelligence as one related by an Italian traveller of some orang-outangs who had had no intercourse with man. When tired of the mountain fruits, or there were no more to be had, they would come down to the sea-shore in search of shell-fish and particularly of oysters. Though apparently reckless in many ways, the monkey tribe have really a good deal of caution, and if, as often happened, the oyster shells were a little open, they were afraid of putting in their fingers lest the shell should suddenly close, as with a spring. To prevent this, the orang-outangs kept the two halves open by means of a stone, so that they could enjoy their oyster to their

THE ORANG DETERMINES TO THROW THE RIVAL MONKEYS OVERBOARD

hearts' content without expecting to be held in a vice at every moment.

Seventy or eighty years ago, the mountain ranges of Cape Colony were infested by swarms of dog-faced baboons, which came, like locusts, to eat and carry away all the ripe fruit from the gardens and orchards. They are very quick, very impudent, and very cunning, and when they lay their plans to rob a garden, they tell off some of the band as sentinels, who give instant warning at the approach of danger. If they are left undisturbed they will not only make an excellent dinner, but will stuff the pouches they have in their cheeks with fruit, to be eaten quietly when they get home.

A traveller by the name of Le Vaillant, who was exploring in South Africa, captured a dog-faced baboon which he called Kees. The two soon became very fond of each other, and were constant companions, for the ape was quick at seeing (or smelling) the presence of wild beasts when the dogs were quite unable to detect them. Le Vaillant turned his greediness and curiosity to account, and never allowed any of his followers to eat strange fruits or plants till Kees had first eaten them, as no ape can bear to pass by food, especially food of a kind he has never seen before. When he threw the fruit away, after merely tasting it, they knew that it was better left alone. Even out hunting, Kees' appetite proved too much for him. He would climb up trees in the hope of finding gum, and dive into hidden places in the rocks where experience had taught him that honey was sometimes to be got. If he could discover neither gum nor honey, he would search for roots, which were the next best thing. There was one in particular which his master enjoyed nearly as much as he did, and when Kees' sharp eyes beheld the leaves, he made all the haste he could to keep it all to himself. First, of course, he had to pull it out of the ground, and that was not so easy. He did not use his hands—this would have taken too long, and besides, the earth was often very hard; but he grasped the plant firmly with his teeth, set his feet tight, and threw back his head with a jerk. If this failed to extract the root, he would then fix his teeth in the stem closer to the ground, and turn head over heels. This was too much for the root, which always came out directly.

Having once got possession of his prize, the next thing was to eat it. He would look carefully round to see where his master was, and would gobble it up more or less fast, according to the distance Le Vaillant was from him, never moving his eyes from the explorer's feet all the while. If, however, his master came on him unexpectedly, he would hastily try to hide the root and pretend that he knew nothing about it; but a light box on the ear soon obliged him to share the morsel with his friend.

In the course of a long day's hunting, Kees, much as he enjoyed the expedition, often got very tired, and used to ride one of the dogs, who, being very good-natured, would carry him for whole hours at a time. As a rule, none of the dogs hated him; indeed he kept them all in order, and if any of them attempted to interfere with him when he was eating, he would adopt the method of his master, and send the intruder away with a box on the ear.

But the biggest and strongest dog of the pack was less good-natured than the rest, and whether from pride or laziness, very much objected to act the part of a beast of burden. So when Kees took it in his head to jump on *him*, he merely stood still and let all the others get well in front of him. Kees could not endure to remain behind anybody, and thumped the dog and pulled his ears to make him go on. But neither thumps nor pulls produced any effect; the dog would not stir. At last, seeing there was no help for it, Kees got down, and both he and the dog raced as fast as they could to join the rest; the dog taking care, however, to keep behind, so that he might run no risk of finding Kees again on his back.

Kees was horribly afraid of snakes, as many human beings are, who have not the least dread of wild animals.

LE VAILLANT AND KEES OUT HUNTING

But even snakes did not fill him with such terror as his own relations—nobody could guess why. At the mere sight of an ape he would scream with fear, and, trembling all over, would creep between the legs of one of the men. After such a shock it was a long while before he was himself again. Being an ape, Kees was of course a terrible thief, and very clever he was at stealing. It was difficult to know how to keep things out of his way, and punishment only made him more cunning. As for hanging up a basket containing milk or any kind of food for which Master Kees had a fancy, it was no good at all! One day, his master had boiled some beans for dinner, and had just put them on his plate, when his attention was attracted by the note of a strange bird just outside his tent. Le Vaillant jumped up, seized his gun and rushed off in search of the bird, which he secured in a few minutes. When he came back to his dinner neither beans nor Kees were to be seen. Of course, Le Vaillant knew what had become of both; but he expected that Kees would appear at tea-time, as he always did when he had been stealing, and seat himself in his usual place with the most innocent face in the world. However, this particular evening nothing was heard of him, and when another whole day passed and no Kees, his master grew very anxious. At last, on the third day, a man, who had been sent to fetch water from the river, reported that he had caught a glimpse of Kees, but that directly the baboon had seen him he had hidden himself in the bushes. On this Le Vaillant called his dogs and went straight to the place where the truant had been hiding, but for a long while could find no trace of the creature. At length he heard a cry—just the sound of reproach that Kees always made when he had been left behind on a hunting expedition, but the animal himself was not visible. His master, in despair, was almost giving up the search, when he suddenly spied the baboon sitting overhead among the thick branches of a tree. Le Vaillant called to him in his friendliest tones, but Kees thought it was only a trap, and would not stir, though he made no attempt to move when his master climbed up after him and coaxed him to come down. When they reached the tent it was quite plain that he remembered his fault, and expected to be punished; but Le Vaillant was too glad to get his pet back to take any further notice. Besides, what would have been the use?

In spite of his penitence—or the shame of having been found out—Kees went on stealing as badly as ever. At least every article of food that disappeared—especially eggs—was always said to have been taken by him, and Le Vaillant determined to discover how far the charge was true. So one day he hid himself near where the hen was kept and waited till her loud cackling told all whom it might concern that she had laid an egg. Kees, who had been sitting patiently on a cart, at once jumped down and ran towards the egg, when his master strolled carelessly towards him. In an instant he stopped, assumed his most innocent air, and balanced himself on his hind legs, as if he had merely come out to see the sun rise. His master pretended not to be aware of the meaning of all this, and turned his back on the bush where the egg lay. Of course the baboon seized it with a bound, and, when Le Vaillant looked round, he was in the very act of swallowing the coveted treasure.

A good whipping followed, but *that* did not save the eggs, so Le Vaillant hit upon another plan. He shut Kees carefully up for a few mornings, while he trained one of the dogs to find the egg and bring it to him without breaking it. Then Kees was let out and Le Vaillant watched with some curiosity to see what would happen. What did happen was this. As soon as the hen began to cackle both ape and dog ran a race to the nest. Each tried to reach the egg first, and in general it was Kees who was the lucky one. If the dog managed to pick it up he brought it straight to his master and laid it in his hand, Kees all the while following, muttering and making faces at him, though he seemed pleased that the dog did not wish to eat the egg himself. If Kees was the victor he bolted with it up the nearest tree, where he ate it in peace, pelting his enemy with the broken shells. Then the dog would return to his master with his tail between his legs.

This was the bad side of Kees; but he had a great many very good qualities. He was an early riser, and when he was up himself he woke the dogs, who held him in great awe, and signed to them to take up their different positions about the tent, which they did without a moment's delay. Then he was devoted to his master, who gives many instances of his loyalty and affection. One day, an officer in fun pretended to strike Le Vaillant, and Kees at the sight became so violent he could hardly be restrained or pacified. The officer, who had not expected the action would make such a deep impression, tried to appease him by offers of fruit, but quite in vain. Never again would the faithful creature have anything to do with the man, and if he caught sight of him ever so far off he would cry and grind his teeth and prepare to fly at him; so that at last, during the officer's stay in the camp, it was necessary to chain him down.

Many, too, were the hardships shared by the pair of friends out hunting, and here, again, Kees' fidelity never failed. The man might sink to the ground worn out with heat and fatigue, parched with thirst, and fainting with hunger, but the monkey never left his side. If there was anywhere within reasonable distance a root or tree that would give them a little relief, Kees would scent it out. Sometimes when found it would have no stalk, so the root could not be extracted in the usual way. Then Kees began to scratch up the hard-baked earth with his claws a painful as well as a slow process—and it was lucky that his master had a hunting-knife with which to come to the rescue. How they would both enjoy that root, when, after so many struggles, they got it at last!

How surprised a traveller would be if, in the course

of his wanderings, he happened to come upon a flock of goats with a baboon for their guardian! Yet this strange sight might have been met with in the land of the Namaquas, about seventy years ago. His master had caught him when a baby, and carefully trained him up to this duty, which he fulfilled as well as any Scotch sheep dog that ever lived. Every morning the baboon drove his charges out to the fields for pasture, and every evening he brought them safe back again, riding always on the

THE BABOON WHO LOOKED AFTER THE GOATS

back of the last, so that he might keep an eye on any stragglers. For wages, he was given the milk of one goat, and he was most particular in keeping to his part of the bargain, and in guarding the others from the hot and thirsty children, who would have been glad of a drink. In the evening his master would give him a little meat for supper.

But the poor fellow was not long left in peace to perform his task. One day, when he was sleeping in the low branches of a tree, he was seen by a leopard, who happened to be wanting a dinner, and after creeping stealthily up, with one bound he landed on the baboon's neck, and there was an end of him.

The Namaquas used to complain that it was difficult to keep a child, for the baboons were sure to steal it, perhaps in revenge for some teasing on the part of the children.

One evening, some little Namaquas were sent out with bows and arrows to play in the woods just outside the village. When it grew dark they all came home again, and it was not until they were close to the huts that they missed the youngest of the party, a boy of five or six, who, being very tired, had lingered behind the rest. Seeing he was alone, a crowd of chattering baboons came swiftly down from their perches in the trees, and seizing the boy in their long arms, carried him off to the mountains.

Next day the whole village turned out as soon as it was light in search of the child, but neither boy nor baboons could be seen anywhere.

For a whole year the parents gave up the boy for lost, when one night a man from another tribe came riding through the village, and mentioned, during the course of conversation, that a long way off he had noticed the trail of baboons, and in the midst the footprints of a child. The villagers set out directly on hearing this news, and when they reached the place described by the hunter, they saw the little boy seated on a high rock, with a big baboon beside him. At the sight of the men the baboon caught up the boy and tried to make off with him, but after a hard chase he was at length surrounded, and forced to give up the child. Far from being pleased at his release from captivity, the boy, who had become quite wild, fought and cried, and even tried to get back to his long-armed friends. He had forgotten, too, how to talk, and it took him some time to pick up his own language again. When at last he had settled down to his old life, the child said that the baboons had been very kind to him, and that seeing he did not like their own favourite food of scorpions and spiders, had given him roots and gum and wild grapes, while, when they came to a spring, they never thought of drinking till he had had his fill. No wonder he missed the good manners of the baboons when he came back to his native Namaquas.

GREYHOUNDS AND THEIR MASTERS

From the very earliest times English people have shown a great love of greyhounds, although as long ago as the days of Canute, no man who was not born a gentleman was allowed to keep one. It must have been their beauty that made them such favourites, and their pretty, caressing ways, for they have not the cleverness of many other kinds of dogs, though their great speed renders them very useful in hunting small game, and even bucks and deer. An old rhyme puts in a few words the qualities that a man would look for in a greyhound, when, as often happened, he wished to send one as a present to a lady, and was anxious to get the best of its kind. It must be

>Headed lyke a snake,
>Neckyed lyke a drake,
>Footyed lyke a catte,
>Taylled lyke a ratte,
>Syded lyke a terne,
>And chyned lyke a herne.

When this prize was laid at the feet of the lady, the giver might ask in return for anything he chose, for women at all times have loved greyhounds, perhaps because there is something that reminds one of a lady in their long necks, small heads, and light delicate figures.

No other breed of dogs has been so often mentioned in history, or has had so many laws made about it. Besides the regulation of King Canute, we find King John taking greyhounds as payment for debts, and accepting

WHEN THIS PRIZE WAS LAID AT THE FEET OF THE LADY, THE GIVER MIGHT ASK IN RETURN FOR ANYTHING HE CHOSE

them as fines. Edward III. kept large numbers of them near his palace at Waltham, not far from Epping Forest, so that they might always be handy when he wished to hunt. The greyhounds have disappeared, but they have left their name behind them, and the place of the royal kennels is still known as the Isle of Dogs.

In the reign of Queen Elizabeth a set of rules for the sport was drawn up by the Duke of Norfolk, and by these rules any doubtful question is still judged. The Queen delighted in coursing, which in those days meant the chasing of deer as well as of hares, and even when she did not care to follow herself, used to sit on some high place and look on from afar. The Stuarts, too, always had greyhounds about them, and of course the courtiers shared their taste; and many are the pictures of the seventeenth century where greyhounds have had, like their mistresses, their portraits painted by the most famous artists.

Froissart, the chronicler, tells a curious story of a greyhound that belonged to Richard II., and was so fond of his master that he did not seem to know there was any one else in the world. It was the only friend the king had when he was imprisoned in the Castle of Flint, and Richard believed that it was clever enough to understand things that had not yet come to pass. 'It was informed me,' says Froissart, ' that Kyng Richard had a grayhound called Mathe, who always waited upon the Kynge, and would know no one else. For whensoever the Kynge did ryde, he that kept the grayhound, did let him lose, and he wolde streyght runne to the Kynge and fawne upon him, and leap with his fore fete upon the Kynge's shoulders. And as the Kynge and the Erle of Derby talked togyder in the courte, the Grayhounde, who was wont to leape upon the Kynge, left the Kynge, and came to the Erle of Derby, Duke of Lancaster, and made to him the same friendly countinuance and chere as he was wont to do to the Kynge. The Duke, who knew not the grayhounde, demanded of the Kynge what the grayhounde would do?

'"Cosyn," quod the Kynge, "it is a great good token to you, and an evil sygne to me."

'"Sir, how know ye that?" quod the Duke.

'"I know it well," quod the Kynge; "the grayhound maketh you chere this day as Kynge of England, as ye shall be, and I shall be deposed; the grayhounde hath this knowledge naturally, therefore take hym to you, he will follow you and forsake me." The Duke understoode well those words, and cheryshed the grayhounde, who would never after follow Kynge Richard, but followed the Duke of Lancaster.'

Among the kings who made friends and pets of greyhounds, we must not forget Frederick the Great, who generally did not waste love upon anybody! He even carried his affection for them so far that he used to take a small variety, known as the Italian greyhound, with him in his campaigns. Once, during the Seven Years' War, he was out inspecting the ground with a view to a battle, when he accidentally got separated from his officers. Hearing a party of Austrians approaching, he picked up his greyhound, and hid under the arch of a bridge that crossed a little stream close by. The enemy, who knew that he was somewhere about, passed the bridge several times in search of him, and Frederick waited in terror, expecting every moment that a bark from his dog would betray him. But the dog seemed to understand how much depended on his silence, and remained perfectly still, till the footsteps had died away. On the death of the little fellow, some time after, he was buried in the dogs' graveyard, belonging to the palace, where each dog has a tombstone, and on it is engraved his name and the good qualities which marked him when alive.

Although, in general, greyhounds are not so ingenious as other dogs, now and then one shows himself surprisingly clever in getting what he has set his mind on.

A story is told of a little Italian greyhound who lived at Bologna in Italy, and was a great favourite with his master. Bologna is a cold place, and greyhounds are often delicate, so a jacket was made for him to wear at night. It was tied on tightly with strings, which were all very well as long as he was lying down in front of the warm stove, but became very troublesome when he wanted to move about and play. So the first thing when he woke he used to run off to anybody in the house who was dressed as early as himself, and jump up on them, and lick their hands till they understood what he was saying, and unfastened his jacket. One day, however, everyone was either ill or busy, and had no time to attend to him, so it occurred to him that, perhaps, if he were to rub himself against the chairs or along the carpet, those tiresome strings would get untied. To his great joy this plan succeeded, and after that he could do without anyone's help. The moment the jacket was off, and the front door open, he rushed across the road to visit another greyhound who lived there with a family, to beg him to come out for a walk. Very often they would spend hours together running races, or playing hide-and-seek between the arches which abound in the streets of Bologna; but he never missed going home to his dinner at twelve o'clock, and again in the evening.

If his friend's front door was not open so early as his own, he would bark loudly to awaken the lazy people; but as they were fond of their beds, they grew very angry, and shied stones to drive him away. Then he stood so close to the door that the stones could not hit him, and barked triumphantly on, till suddenly the door was flung open, and a man appeared with a whip. The dog could not think of any way to get the better of the whip, so he walked off to consider what was to be done.

A few days later, he went back to the door and waited quietly till it was opened; but the people had taken a dislike to him, on account of all the trouble he had given them; and as soon as they saw him, drove him away. After that he did not go near them for some time, and when he paid his next visit, placed himself out of reach both of stones and whip, and then barked away as loudly as ever.

He had nearly barked himself hoarse out of pure revenge, when a boy came to the house, seized the knocker, and let it fall again. Then, to the surprise of the dog, the door was opened, and the boy entered the house. When he had recovered a little from his astonishment, he crept slowly along the wall, till he reached the very place where the boy had stood. Then he jumped up to try to catch hold of the knocker, but it was high up, and he had to jump a great many times before he managed to catch it between his teeth. It fell with a great bang, and some one called out, 'Who's there?' and, as the dog was silent, came to the door and threw it open. In flew the dog, and ran straight to his friend, whom he had not seen for so long, and received a warm welcome. The family were so much amused at his cleverness that this time they let him stay, and whenever his morning 'rat-tat' was heard, it was a race between the children as to who should answer it.

At the time when I am now writing (Dec. 20, 1897), there is an account in the papers of the rescue of a dog from a ledge on one of the highest and steepest cliffs in Dover. Some boys looking down from the top, saw the little liver-coloured creature lying, with a lady's hat beside it, more than two hundred feet below, and told the police, who said it was quite impossible to get at her. A week passed and the dog was still there, and the boys could stand it no longer. With the help of a man named Joys, they drove footholes into the cliff from beneath, and managed to reach the little spaniel, after a dangerous climb of about a hundred feet, while the coast-guard let down ropes from above, and hauled them all up together. The poor little thing was terribly weak from her long fast, and the nights and days she had passed on the rock, and Joys, who carried her home, feared that it was too late to bring her back to life. But the careful nursing of himself and his wife has done wonders, and she is as strong as ever she was.

How she ever got to that ledge, or what has become of her mistress, is still a mystery, and perhaps it always will be. Whether the hat was blown over the cliff, and the poor lady, trying to catch it, overbalanced herself and was carried out to sea, while the hat remained stuck on the rocks, and was followed by the dog, nobody knows, and most likely never will. But the dog has become quite a famous person, and several offers have been made to buy her from her kind hosts, so that she is quite sure to be petted as a heroine to the end of her days.

THE GREAT FATHER AND SNAKES' WAYS

PROBABLY no wild beast that ever lived has caused such deadly terror to so many people as that inspired by snakes of all kinds. With a lion or a tiger a man feels that he knows pretty well what the creature will do, and how he must defend himself. The animal springs, and bounds, and bites, and men can spring, and bound, and bite, too, if they want to, though not so far, or so well. As regards a snake it is quite different. His ways are not our ways; his method of getting along is totally unlike ours; he does not display a great row of gleaming teeth to frighten you and tear your flesh; his head darts at you like lightning, and is as quickly withdrawn; but in that instant, unless strong remedies are at hand, your death-blow has been struck.

It is this sense of mystery and strangeness that hangs round serpents which makes them the object of such dread, though of course there are many kinds which are perfectly harmless. Dark tales, too, are told of their strength and power of fascination, by which their victims are not only prevented from making their escape, but even forced to advance towards their fate. As a rule, however, snakes, unless they are very hungry, only attack in self-defence, and act on the principle that 'if you do not hurt me, I will not hurt you.' Still, without meaning to hurt them, they sometimes look so like a dead branch that they get trodden upon; then woe be to the creature who has roused them from sleep!

Snakes are usually largest and most dangerous in hot countries, and Dr. Livingstone tells of one in South Africa that is over eight feet long, and has an immense amount of poison in its fangs at once. He has seen it attacked by a herd of dogs, and all four of them stung to death. Of course the poison gets weaker the oftener it is used, and while the first dog dies at once, and the second in five minutes, the one who has received it last may linger for some hours. He mentions a snake that he saw killed, which contained in itself such vast supplies of poisonous fluid that, even after its head was cut off, the fangs continued to drop it for many hours. This particular snake has, according to the natives, a horrid trick of spitting its poison straight into the eyes, with the result of blinding its victim; but we are not told whether it can cause death without a distinct bite.

In cold countries snakes generally seek out a warm place when the air begins to grow chilly, and stay there till the summer comes back. Long ago, a strange thing occurred in the house of an English gentleman living in the country, with a servant who had been with him from a boy. Now this servant, says the chronicler, 'grew very lame and feeble in his legs, and thinking he could never be warm in his bed, did multiply his clothes, and covered himself more and more, but all in vain, till at length he was not able to go about, neither could any skill of physician find out the cause.

'It happened on a day as his master leaned at his parlour window, he saw a great snake slide along the house side, and to creep into the chamber of this lame man, then lying in his bed (as I remember) for he lay in a low chamber, directly against the parlour window aforesaid. The gentleman, desirous to see the issue, and what the snake would do in the chamber, followed, and looked into the chamber by the window; where he espyed the snake to slide up into the bed-straw, by some way open in the bottom of the bed, which was of old boards.

Straightway his heart rising thereat, he called two or three of his servants, and told them what he had seen, bidding them go and take their rapiers, and kill the said snake. The serving men came first and removed the lame man (as I remember) and then the one of them turned up the bed, and the other two the straw, their master standing without at the hole, whereinto the said snake had entered into the chamber. The bed was no sooner turned up, and the rapier thrust into the straw, but there issued forth five or six great snakes that were lodged within. Then the serving men, bestirring themselves, soon despatched them, and cast them out of doors dead. Afterward the lame man's legs recovered, and became as strong as ever they were; whereby did evidently appear the coldness of these snakes or serpents which coming close to his legs every night, did so benumb them as he could not go.'

It is often supposed that snakes are unable to make any sound but the terrible hiss they utter when they are angry; but there is one African kind that imitates the cry of a kid so exactly that it is impossible to tell one from the other, and many is the animal which, thinking to find a goat, has fallen into the trap set for it by the serpent!

Another species, that has a sort of voice in its tail, (as well as one in its throat), is the rattle-snake. The famous 'rattle' that it sets up whenever it sees an enemy approaching comes from the shaking together of a loose, horny, jointed substance at the end of its tail, and when a man or animal hears this he knows what awaits him, and can get out of the way if he chooses. If he does *not* choose, but prefers to attack the rattle-snake, which twists itself straight up in a wreath of many coils, its eyes gleaming from the centre, he runs the risk of a speedy and painful death. The teeth that the snake uses as his weapon of defence are two very small and sharp ones in his upper jaw, which have each a little bag at its root, containing the poisonous, greenish fluid. This fluid spreads itself through the blood with wonderful quickness, and a chill instantly strikes the whole body. Then the spot

THE SNAKES FOUND IN THE LAME MAN'S BED

where the poison entered begins to swell and get discoloured, and soon the whole frame shows the same signs. The poison takes effect most quickly in hot weather, or if the wound is just above the heel; but, strange to say, it never affects the wild hog, who can even eat rattle-snakes without suffering.

Luckily, certain remedies are known to the Indians for the bite of one of the deadliest of all serpents, the rattle-snake, and the surest of these is a small kind of plantain which, when rubbed on the wound and swallowed, gradually destroys the poison in the blood. Now-a-days, too, people are given strong doses of whisky or ammonia, which act in the same way, and they are kept walking up and down for many hours. If they are once allowed to fall asleep they never wake again. Still, whatever remedy may be used, when the time of year comes round in which the man was bitten, he will, we are told, feel some return of the symptoms to the end of his life.

A traveller in North America, in the middle of the last century, says that in the neighbourhood of the Fox River he found an immense number of rattle-snakes hiding in the grass which covered a sort of swamp. One of these snakes had been captured by an Indian, who managed to tame it, and carried it about with him everywhere in a box, calling it his 'Great Father.' The man and the snake had wandered about together for many summers, when they were met by a French trader, who found the Indian making ready to start for his winter hunting-grounds. He and the Indian soon became friends, and one day the Frenchman was much surprised to see the Redskin put the box containing his Great Father on the ground, and, pushing back the lid, tell the snake, as he did so, that he was to meet him at that very place the following May. The Frenchman laughed when he heard the Indian's words, and said that as this was only October he hardly thought it likely that the Great Father would remember so long. However, the Indian was so certain of the snake's affection that he offered to pay the Frenchman two gallons of rum should the Great Father not turn up at the time appointed.

When May came round, the French trader set out for the place, and found the Indian there before him, the box in his hand. He laid it on the ground, and called to the Great Father, but the Great Father never came. After waiting some time, he acknowledged that he had lost his bet; but he still felt so certain of the snake's return that he offered to pay, not two gallons, but four, if the Great Father did not appear within two days. The Frenchman agreed to this, and the second morning saw both men on the spot.

Some hours passed, and the Frenchman was already counting on the gallons of rum, when, about one o'clock, a motion was seen in the grass, and the Great Father glided rapidly towards them. He made straight for his box, as a man would make for his house, and without waiting to be told, he curled himself up snugly inside, and the door was shut on him.

ELEPHANT SHOOTING

It was early in the month of November that Baker went down to the last cataracts of the White Nile, about six miles to the south of his camp.

The country was everywhere very rich, and covered with villages, and the people were very friendly, and ready to give the new comers all they wanted in the way of food.

One day a troop of Baker's soldiers had been sent to some distance to fetch corn, and while their commander was quietly sitting smoking on the deck of the boat the leader of the party came galloping back to say that a herd of elephants was coming up from the west of the river.

Baker did not pay much attention to this news, as he expected that the moment the herd caught sight of the people, who had from curiosity climbed on the rocks or squatted on the roofs of the huts, they would turn off in some other direction. But the elephants did nothing of the sort.

On they came, eleven in number, swinging their trunks and flapping their ears, not seeing or not heeding the crowd of boats and people.

When they arrived within about four hundred yards of the river, Baker mounted his horse— Greedy Gray—first telling his servant Sulciman to send on his two elephant rifles, with plenty of powder and ball. He then posted some of his men, dressed in red shirts, on the low hills close by, with orders to come down behind the elephants so as to prevent their turning. This done he galloped up the slope, taking care to keep well above the herd.

By this time the elephants had reached the river bank, and at sight of the grey horse they stopped suspiciously and stood closer together. While they were standing thus the men came down from the slopes and formed a long line, surrounding them on that side. The elephants remained quiet, though they still flicked their ears, with the boats in front and the river behind them. Here the stream was broken, about a hundred yards from the shore, by an island, with a steep bank of hard earth. Before a shot could be fired they had swum across and gained the island, but then their progress was stopped. The banks were fully six feet high, and the river was too deep below to give them a footing. The only thing they could do was to pull down the bank with their trunks and tusks, so as to form a slope for them to pass up, and they at once set about it.

Hard as the whole eleven worked, it took some time; and Baker, standing on the shore, watched closely for the moment when one of them should turn round, for it was difficult to shoot with any certainty from such a distance. However, he did fire a few bullets in among them, which, though they did no real damage, bothered the elephants a good deal, and caused them, in their confusion, to tumble over each other. But by this time part of the bank had given way under their hard labour, and they were enabled to get some sort of footing above the water, so that more of their bodies were exposed to view. At last, with a prodigious amount of tumbling and struggling, one large animal reared itself half out of the river, and received a ball behind its shoulder. It fell over into the stream, which swept it quite near to where Baker was standing, so that it was easy to put a ball right into the brain.

When his rifle was loaded for the second shot an elephant had scrambled right to the top of the bank, and gave an excellent mark, which Baker did not fail to take advantage of. This animal was killed on the spot, and like the other rolled into the river, and boats were sent at once down the stream to tow them both back.

But luckily for the rest of the herd there was no more ammunition left, so the elephants were allowed to climb up the bank without any more disturbance. They took counsel together as to what was best to be done, and at length agreed it was safer to cross to the further side. A few stray shots from a field gun hastened their movements, though the shells burst without touching them, and the whole nine were soon out of reach on the eastern shore.

As for the two which had been killed, the current was so strong that the boats sent after their bodies had to go two miles before they came up with them. Unlike a hippopotamus, which sinks for two hours after he is dead, the elephant always floats, for he is like a great football, on which two or three people can stand. The hippopotamus, on the other hand, is solid all through, and his skin is far thicker and heavier than the elephant's.

The two heads and the tusks were all that Baker wanted, so he was pleased to gratify the villagers who crowded round begging for the meat, which they are very fond of. Hundreds of them came flocking, while some of the tribesmen, who had shown themselves unfriendly, looked on in disgust, watching the preparations for the feast. They were very much awed, too, by the way in which the animals had been killed, and dreading, like all savages, anything they did not understand, they at once sent messengers to beg for peace, which was cheerfully granted them.

Elephants are very particular what they eat, and prefer roots, bulbs, or the branches of trees containing sweet, gummy juice—like mimosa—to anything else. In their turn their flesh is much prized by the people, partly on account of the fat, which is not only eaten but

BAKER SHOOTING THE ELEPHANTS AT THE ISLAND

smeared over their bodies. The ivory tusks are, of course, used as an article of trade.

Along the course of the Zambesi river elephants are to be found in vast herds, or *were* to be found, sixty years ago, when Livingstone explored that country. One way of killing them is to make platforms high up in the trees, under which the elephants must pass. As soon as the animal is right under the trees a man aims a spear, measuring four or five feet, with a sharp blade twenty inches long, straight at the elephant's ribs, and a well-directed blow causes death very soon. Sometimes they use instead of this a spear fixed to a beam of wood and hung on a dangling cord tied to a tree. The head of the spear is poisoned, and when the animal treads on the cord the spear wounds him in the foot, and he dies in a few hours.

In these regions men are forced to do their hunting on foot, for horses fall victims to the terrible tsetse fly, from whose bite neither ox, horse, nor dog ever recovers, though it never touches either wild animals or men. It is, therefore, very difficult to kill an elephant with one shot placed in the brain, as is done in countries where horses can be used, and, besides, the climate makes hunting a very tiring sport, and only fit for very strong men.

In 1850 a friend of Livingstone's, named Oswell, was tracking an elephant along the banks of a river, and saw him with disgust take refuge in a thicket of thorny bushes, which did not hurt his hard skin, but were very unpleasant to a white man. Here the country was comparatively free from tsetse, so Oswell was riding, and at once put his horse into the narrow path, forcing his way as well as he could through the dense branches. When he was well into the midst of the tangle, keeping his eye steadily fixed on the elephant's tail, the creature turned suddenly round and charged. Oswell tried to break away in another direction, but found it was hopeless, and in leaping from his horse caught his foot in a branch and fell to the ground, touching in his fall his horse's side with his spur. The animal plunged and bolted, and before Oswell could rise the elephant was upon him. He expected every second to be crushed by the weight of its enormous feet, but the elephant, in its wild rush, had not seen his fall, and passed him by, positively placing his foot between Oswell's legs, which he had instinctively parted. Few men have had such a narrow escape, and indeed he had been saved from more than he knew, for these thorn bushes cut like knives, and few horses will face them.

It is the custom of the Bechuanas to dig pits for the animals to fall into, after the manner of the Scotch at Bannockburn. The shape they have found to answer their purpose best is a kind of long square, seven or eight feet deep, but only one foot wide at the bottom, while the breadth at the top is at least three or four feet. When finished the pits are carefully covered up, and all traces of disturbance removed by a sort of framework of reeds and grass, held together by sand. In leaving the banks of a river, where they often go at night to drink and wash themselves, an old elephant will be placed in front so as to examine the ground, lest pitfalls should beset their track. And if sometimes, in spite of all the care of the leader, a young and foolish creature blunders into a hole, the strongest among them will join together and by means of their tusks and trunks will drag him out of his death trap.

Indeed, elephants, like many other animals, have strong affections, and will often attach themselves to one of their own herd, defending it from all dangers, as the following story will show.

Colonel Gordon Cumming was hunting elephants in the country north of the Limpopo river, and they frequently led him a long dance, for if they suspect a man's presence in their neighbourhood they will go miles to get out of his way. They even seem somehow to tell one another, for if one has been shot all the herds in the district hear about it, and in a day or two there is not an elephant to be found anywhere. When the sun is hot they will shelter during the day in dense jungles of 'wait-a-bit' thorns, only coming out every third or fourth day to drink and wash in some pool or river, very often thirty or

OSWELL'S NARROW ESCAPE

forty miles away. This done they go back to a sheltered place and lie comfortably down to sleep, with their backs to an ant-hill, which shows an odd taste in beds.

Well, early one morning Colonel Gordon Cumming left the hole in which he had been sleeping, and climbed up a high rock to see if there was any chance of an elephant. Yes, sure enough, there were nine or ten huge creatures having their breakfast not a quarter of a mile away. It seems strange that so many people can see no animal, however harmless, without wishing to kill it; but Colonel Gordon Cumming had travelled thousands of miles for no other reason, and his heart beat high. He quickly clambered down from his rock to warn his men to keep quiet and out of sight, and sent back to the camp for a fresh horse, his dogs, and his big rifle. Then he returned to his watch-tower to make out the lie of the land.

The first herd he knew, from the size of the beasts, to be made up entirely of females, with some young ones following closely at their heels; but further away was another troop, consisting of five males, also grazing quietly. These he resolved to leave till the horses and dogs came up, and to hunt the others on foot.

Very cautiously he moved along the rocky ridge where the females were feeding on the young branches of the trees, till he got within a hundred yards of them. As the wind was blowing straight at him the elephants scented nothing, but continued to approach, munching as they walked. The sportsman picked out the largest and fired. The elephant uttered a cry of surprise more than of pain, and turned sharp round, receiving as she did so a second ball in the shoulder. Growling and muttering, the whole herd set off at a sharp trot northward, flapping their huge dangling ears as they went, the wounded female bringing up the rear with a friend by its side. When they reached a clump of trees they stopped, and not having scented man they thought they were safe. Meantime the horses and dogs had come up, and the hunters rode slowly towards the grove.

They had not gone far when the elephants caught sight of them, and started off afresh. But the poor wounded one could not keep up with the rest, and was easily cut off. Gordon Cumming dismounted, and, throwing his bridle over one arm, tried to aim steadily at the elephant. He found this, however, almost impossible to do; his horse Colesberg was in mortal terror of the huge, strange creature, and plunged wildly. A shot was at length fired, but without much result, and the noise at such close quarters ended by upsetting Colesberg's nerves completely. In vain his master attempted to get near enough to jump on his back; Colesberg only plunged and reared and swung round towards the wounded elephant. At this moment a loud trumpeting noise was heard from behind, and out from the trees came a stone-deaf old dog, followed, unknown to himself, by the friend of the wounded elephant, who had come to the help of his comrade. The men looked on from afar; but, less loyal or brave than the elephant, they did nothing, and Colonel Gordon Cumming's hunting days would have ended there and then had it not been for the dogs who yapped at the knees of the elephants, and took off their attention—for elephants are horribly afraid of dogs. When their trunks were almost touching him he managed, goaded by the danger, to spring into the saddle, and dashed off to where the men were standing for a second rifle. Then, aiming as well as his frightened steed would let him, he soon ended the sufferings of his first victim, which fell to the ground, bringing down a huge tree in her fall.

Her friend, seeing the case was hopeless, charged straight at the murderer, who was forced to fly for several hundred yards before he could contrive to get a shot. At last he was able to turn and place a ball in her shoulder, when, evidently hard hit, she gave it up and made for cover.

Some old writers have left us very curious stories of the elephants which were first seen in Europe in the wars of Pyrrhus with Rome. 'The beast which hath between its eyes a serpent for a hand,' was much used in battles in those days, and when steady and well-trained, was most useful, both in charging the enemy, and in carrying a kind of fort filled with light armed men on its back. In the wars between Carthage and Rome, Hannibal is said to have ranged his elephants in the front of his lines, to break the shock, and to trample down the advancing foe. But in the end, the Romans got accustomed to these tactics, and learned how to foil them.

A number of these tales—not always true or even likely—were collected about two hundred and fifty years ago by a man named Topsel, who published them in a book called 'The History of Four-footed Beasts and Serpents,' illustrated with some very funny pictures. Topsel assures us that in their wild state old elephants are cared for by the young ones, who gather food for them and fight for them when they are not able to fight for themselves; and that when they are dead, green boughs are laid over them by the rest of the herd. He further declares that they have been known to pull darts and spears out of each other's bodies, and that when Porus, king of the country beyond the Indus, was defeated by Alexander the Great, his favourite elephant drew the javelins out of his wounds with his trunk, and then knelt down very gently, so that if the king was still alive, he might not be shaken. Topsel does not tell us whether it was Porus, or another Indian king, who had a bodyguard of elephants, which were trained to watch him by turns while he was asleep, and never failed to appear at their appointed hours, like sailors on board ship.

One story he quotes from Arrian the writer, of an Indian who had brought up a white elephant from the time it was a little creature, and loved it dearly. Now white elephants are greatly valued in many countries; indeed, in Siam, they take rank immediately after the king, and before the heir to the throne; and the king of that part of India, hearing of the white elephant, sent to the man and demanded it should be given him as a present. The Indian could not bear the thought

HANNIBAL'S ELEPHANTS

of parting with his elephant, which he had brought up and taught for so many years, till it was almost like his own child, and in the middle of the night he mounted its back, and they both fled away into a desert place.

When the king heard what the man had done he was very wroth, and sent messengers to take the elephant, and to bring its master into his own presence, so that he might receive the punishment due to his disobedience.

The Indian saw them coming, and climbed with his elephant up a steep rock, only answering their summons to give himself up by throwing stones at their heads, and the elephant followed his example. At length, some of the men stole round from behind, and seizing the Indian threw him on the ground. At this the elephant waxed so furious that it charged them madly, catching up some in its trunk and dashing them to the earth again, and trampling others under its great feet. The men at the back, seeing the fate of their foremost comrades, fled away in terror from the enraged elephant, who then, stooping over its unconscious master, raised him gently in its trunk, and carried him away to a safe place.

HYENAS AND CHILDREN

Long ago, travellers used to think that hyenas had a kind of magic about them, by which they could force their prey to stand in one place till they were ready to fall upon it. It was enough for the hyena to walk round an animal three times to make it as helpless as a bird in the power of a snake. Of course it may not always have been easy to get the creature to remain stock still while the hyena was performing this ceremony— for nothing less than three complete turns would induce the spell to work —but that does not seem to have occurred to the old writers. In the case of a man, he must be most careful, if he ever met a hyena, not to allow him to pass on the right side, for, if he did, he would be certain to fall senseless off his horse before he had ridden very far.

All sorts of charms were considered necessary to preserve men against the wiles of a hyena, and curiously enough, the beast's own skin was held to contain a spell. A hyena's skin hung up on a gate or fence would ensure that the fruit trees within should be proof against either hail or lightning. No darts could pierce the man who went into battle with the skin of a hyena wrapped about him, and any farmer, anxious to increase his crop, had only to place his seed in a hyena bag, for his land to bring forth a three-fold quantity.

Travellers in these times do not put quite so much faith in the power of the hyena, dead or alive; but they quite agree that, like most cowardly beasts, he is very cunning. One of his favourite tricks is suddenly to pop up his big, bristly body in the midst of the grass where a herd of cattle are feeding, and to frighten them into running away. If they fly, he follows and bites the hindermost animals; but if they make a stand, he seldom or never fights.

He may always be found prowling about African battle-fields, after the dead bodies, or hiding amidst the ruins of a deserted city, waiting for some unsuspecting creature to come past; but his greed is equal to his cowardice, and he infinitely prefers human flesh to any other, when he can get it without danger to himself. In some parts of Africa hyenas will actually pass by a calf which is tied up inside the dwelling-house, and will take up a sleeping child from its mother's side, moving it so gently that neither child nor mother wakes.

This horrible event happened more than once in the family of a man named Dassa, who lived in the land of the Kaffirs. One night, when all were sleeping soundly, a great hyena (or wolf, as it is called in those parts) stole softly into the house, and picking up a little boy, made for the door. Luckily, however, the child woke before it was beyond help, and its cries brought its father to the rescue. The hyena dropped his prey, and the boy escaped, thankful to get off with a torn cheek.

The next night the father lit a bigger fire than before – very few animals will venture past a fire—and lay down to sleep with his longest spear in his hand. But the hyena knew better than to come back so soon, and if he peeped in longingly, nobody was any the wiser.

Several nights passed; the boy's cheek was almost well, and the fright nearly forgotten. But one morning, when they all woke, another brother was nowhere to be seen. They searched high and low, with beating hearts, and at length they came on the trail of a hyena, which they followed up. Then at last the father came on something which he recognised as having belonged to the boy, but nothing more was ever heard of the little fellow himself. The terrible creature had managed to move him so softly, that the child had never opened his eyes till it was too late, and then he saw the face of his deadly enemy above him. It really seems almost as if hyenas had indeed the power of casting a spell over their prey.

Made bold by this success, the hyena prowled nightly round the house till he thought he could safely venture in. This time, the child lying nearest the door was a boy of about ten, and therefore not so easy to deal with as the other two. In spite of this, the monster managed to get him outside the hut, and then dropping him for a moment, seized him again by the shoulder. The child, now fully awake, gave him such a blow on the nose that the hyena let go his shoulder, but grasping his collar-bone firmly with his teeth, broke it in two. The boy still hit out, though his right arm was disabled, and again the hyena shifted his hold to the thigh, and ran off with his victim just as the father, roused from sleep by the boy's cries, rushed to the spot. For a quarter of a mile the chase continued, and then a fierce blow forced the animal to drop the boy. When brought to the camp of the Englishmen to be doctored, his thigh was found to be half-bitten through, though fortunately the bone was not broken. Every possible care was given to him, and in a few months his leg was quite cured.

A little girl, two years younger, did not escape so easily. One very hot morning, she was lying under the shade of a tree, when four hyenas suddenly appeared before her, and carried her off between them. One took hold of her head, a second seized her shoulder, and the other two grasped her thighs. The child's shrieks brought the village people flying to her help, and they soon managed to beat off the hyenas, but not before the child was dreadfully injured. Savage nations have not much patience with sick or deformed people, even when part of their own family; and after trying all their medicines on her for a few days without making her any better, they got tired of the whole affair. A choice was given the poor little thing, either to be put to death at once, or, if she preferred it, to be turned loose in the forests, and there to run her chance of being eaten by wild beasts, or dying of starvation.

It did not take the little girl long to make up her mind, neither did she waste any time in weeping over her dreadful fate. Although only eight years old, she had heard of the fame and the kindness of the white men, and she at once determined to go in search of their station. What a terrible journey that must have been! The station was many miles away, and to reach it, the child, badly injured as she was, and still suffering from the shock of the attack, had to pass through woods haunted by the most savage beasts, and to climb through deep glens, where an enemy might be lurking behind every rock. But somehow or other she did it, and arrived at the station in a fearful condition of pain and hunger, covered with fourteen large wounds from the teeth of the hyenas. At first it seemed impossible that she could live, but, wonderful to say, in the end, she not only recovered from her injuries, but bore hardly any signs of them, except some scars.

In spite of his ugly, ill-shapen form, few animals are quicker of movement than a hyena, and, cowardly though they are, their skill in dodging often enables them to get the better of their enemies. When Mr. Selous was on one of his expeditions in Mashonaland, the camp was disturbed for several nights by the knowledge that a hyena was prowling round, in the fond hope of catching them napping. He had not, however, shown as much cunning as usual, for the moon was still bright, and it was easy for the dogs to stop his proceedings.

At length a night came when the moon did not rise till ten, and, as near the equator it always gets dark early, it was necessary to shut up the camp at sunset to defend it against wild beasts. So the waggon was, as usual, placed in the middle, and the horses tied up just beyond, with their maize porridge cooling beside them on the hide of a freshly-killed eland bull. A few yards away were a circle of big fires, with thirty or forty natives talking and laughing over their supper.

Suddenly, in the very midst of the group, appeared the gaunt form of a hyena, with its sides looking as if they had been flattened by a spade. It seized the skin, and was lost in the darkness, before any of the men had recovered from their surprise. Indeed, the whole thing hardly lasted longer than a flash of lightning. In a moment, however, when they had recovered their senses, they were all after it, dogs as well as men, lighted by bundles of burning grass by way of torches. The trail was easily found, as it had to drag the huge eland skin, weighing at least forty pounds, in its mouth, but it was already across the stream, three hundred yards away, before the dogs came up. Then it dropped the skin at once, without attempting to show fight, and galloped off as fast as its legs would carry it.

But they all knew the ways of hyenas well enough to be sure that this one was certain to return again before very long. So the dogs were tied up, and as there was still plenty of time before the moon rose, Selous took his rifle and waited under a bush outside the camp. After some time he fancied he saw something coming towards him, and when the creature was quite close he fired. It was too dark to tell clearly what had happened, but it seemed as if something fell, and then got up and walked off. Shouting for the dogs to be unfastened and for the Kaffirs to bring torches, Selous made ready to follow, and the hyena was tracked to some long grass a hundred yards away. It managed to beat off the attacks of the dogs, and reached the river, where it stood in a pool till an assegai from a Kaffir put an end to it, much to the joy of the natives, for the hyena was a well-known robber, and many were the goats and cattle that it had stolen for dinner.[1]

[1] Steedman's *Wanderings*, and Selous' *Travels passim*.

A FIGHT WITH A HIPPOPOTAMUS

The great White Nile river, which flows north, out of Lake Victoria Nyanza, and joins the Blue Nile at Khartoum, is full of hippopotami, who lie concealed in grassy swamps on the river bank by day, and come out to play in the cool of the evening. In many places this river is choked up by mud and vegetation, so that very often the water is not more than five or six feet deep, therefore only small boats can float easily. Under these circumstances a huge heavy beast, like the hippopotamus (which means in Greek 'river-horse'), can do great damage, and travellers and explorers have many tales to tell of their narrow escapes.

Nobody had more adventures with these troublesome animals than Sir Samuel Baker, when, thirty years ago, he set out from Khartoum on his journey south. Sometimes the hippopotamus would be seen leaving his grassy bed, where he had been sleeping during the long hot day, his hard skin preserving him from the flies which are the pests of those countries. But more often his presence would be guessed by an agitation on the surface of the stream, and a loud snorting noise, and then his ugly, shapeless head would be thrust out.

With such a thick hide to deal with, Sir Samuel preferred, in his encounter with a hippopotamus, to use a weapon more certain than an ordinary bullet. He liked to allow the animal to get within thirty yards of him, and then to take accurate aim right under his eye, where the bone is thinnest, and the brain can most easily be reached. The bullet he employed was of a very deadly kind, being really an explosive shell in the form of an iron bottle, filled with strong gunpowder, and fitting into a two-ounce rifle. This shell did great execution, and produced instant death.

One night the party lay encamped by the side of a lake, and a small boat was moored to a grass-covered mud bank, close to the larger vessel. This boat, used as a larder, was full of hippopotamus flesh, which the men considered a great treat, and did not seem to find too tough for their strong white teeth.

After dinner was over, and the mosquito curtains hung up, the natives dropped off to bed one by one, and soon all was quiet, except for the sentry's steady tramp. Considering the latitude the night was cold, and wrapped in blankets everyone slept more soundly than usual. Suddenly Sir Samuel, who lay on a sofa on the poop-deck, was roused by the sounds of loud snorting and splashing just below him, and by the light of the moon saw a huge hippopotamus making ready to assault the little ship. Calling to his servant, Suleiman, to bring a rifle, Baker made his way to the main deck, where the rest of the crew were sleeping; but the whole place was such a mass of interlaced mosquito strings, that it was very difficult to steer between them so as to wake the men. Meanwhile the hippopotamus had not wasted his time. He had sunk the larder boat, and crushed the little 'dingey' alongside as if it had been a walnut, and was now gathering himself together for the larger vessel, caring nothing for the noises made by the natives in the hope of frightening him away. As for Suleiman, he was in such mortal terror that he never remembered that the gun he had brought was unloaded, and that he had forgotten the charge.

Thrusting him hastily on one side, his master dashed

into the cabin, where ammunition and loaded rifles were always kept; but for a few minutes the commotion raised in the water by the furious beast was so great that no aim could be taken with certainty. Then the shell, which had always proved so deadly, was sent at him, but produced no effect except to make him still more wild, and

HOW THE HIPPOPOTAMUS ATTACKED THE BOAT

the boat rocked wildly about as if blown by a hurricane. Several other shells were fired at him, but for a long time he gave no sign that any of them had touched him, then he slowly drew himself out of the water, and lay still snorting in the swampy grass. Taking for granted, rather rashly, that he had received his death-blow, Baker gave the order for everyone to return to bed, as the danger was past.

But he had rejoiced too soon. In half an hour that fearful splash was heard again, and with a rush the creature made for the boat. A bullet in his head stopped his career just as he was upon it, and rolling and kicking, apparently in his last agony, he was carried down stream.

After he had floated about fifty yards he suddenly, to the surprise of those who were watching him, pulled himself together, and returned slowly along the river bank, which lay in dense shadow. The boat's crew waited with their ears at full cock for some time longer, and then decided that the beast had had enough, and that they might go back to bed for the third time. Baker followed their example, but kept the gun close beside him.

Unlike his men, he did not feel inclined to sleep, and it was not long before everyone was again on his feet, watching the enemy, who was splashing heavily across the river so as to get a better chance for a rush. Now was the opportunity for aiming at the shoulder, and as the animal turned and his body was exposed, Baker lodged a ball in his heart. This time he really was dead, and tumbled into the river.

Then they all went to bed again. Next morning they examined his body—which was covered with scars from the tusks of his own species—for the fury of his onslaught really looked more like madness than anything else. The bullets had broken one of his jaws and cut through his nose, but nothing except death could stop him from fighting. As for the dingey, he had simply bitten out a piece of its side, and would doubtless have done the same to the larger vessel if he had been suffered actually to touch it.

KANNY, THE KANGAROO

A WRITER in *Chambers' Journal*, more than twenty years ago, tells an interesting story about a pet kangaroo that he and his sisters had for a playmate. How she came into the family he does not say. Perhaps some sailor uncle or cousin brought her from Australia; but, at any rate, there she was, and dearly the children loved her.

To begin with, she was so pretty, tall, and slight—she measured quite five feet when standing up—with a small head, large eyes, and soft silky skin. Her tail, which she used both as a whip and as a means of expressing her feelings, was long and powerful, and with her two little hands she helped herself at meals in the most delicate and polite manner. And then, how she could jump! The flight of stairs she cleared at a bound, with an ease no boy ever managed to imitate; and as for the big hall, four skips brought her from one end to the other. The cats, who had been rather pleased with their own leaping performances before Kanny came, treated her coldly, and not very civilly; when she bounded into the room where they were all comfortably seated on the best chairs, they rose as one cat, and put their tails up and their ears down. Kanny did not understand the language of cats —it was only quite lately she had made the acquaintance of any—and stared at them with wonder, and when the cats found it was no use being rude, they became polite, and at last grew quite fond of Kanny, who never tried to take liberties with them, though she *was* so big. But to the end they never could bear to see Kanny help herself first at dinner, and growled and snarled when she put her paws into the dish.

Kanny's favourite dinner was rabbit bones, and this taste was shared by the cats, but in general they considered that, in the matter of food, she was not to be depended upon. Fancy any sensible creature liking tea,

THE NEW ARRIVAL

when it could get good milk, and sometimes—say on birthdays—cream! What *could* she see in all those horrid pink and yellow things that the children called 'bull's eyes,' and 'lollipops'? and surely she must be mad to get so excited over those hard white fruits that were said to be almonds. But Kanny paid no heed to these remarks and scornful glances, and ate thankfully all the sweets that the children gave her. Indeed, almost the only time she was ever out of temper was when anyone forgot to put sugar in her tea!

One day, when the children came in from their walk, they went as usual to find Kanny in her own particular apartments an outhouse, which had been fitted up for her when she first arrived. Instead of bounding to meet them directly she heard their voices, as was her usual habit, Kanny waited for them in her own drawing-room, quite like a lady. As the children ran up to her they suddenly stopped short, for out of Kanny's pouch two little black eyes, and two little skinny hands were peeping. Oh, how happy the children were with the new baby, and what care they and Kanny took of it. Other people might say it was lean, and ill-made, and ugly, but *they* knew it was the most beautiful creature in the world. For a few weeks nothing was thought of or talked of but the baby kangaroo; then frost came and cold winds blew and one day the poor little baby was found dead in its cradle.

The children did their best to comfort Kanny, and brought her all the sweet things they could think of, and by-and-by she began to play again. Her favourite prank was to jump on top of the great walls seven feet high, which shut *in* the fruit garden and shut *out* the children, and then spring right down among the bushes where her favourite currants and cherries grew. But Kanny's appetite was a good one, and like all people who are fond of eating, she enjoyed trying experiments. One morning, when she had had as much fruit as she wanted, she leaped on top of the wall which at that end opened out upon a lane, where some workmen were busy making a door into another garden. Her movements were as silent as they were rapid, and when the carpenters suddenly looked up and saw this strange creature standing before them, they flung down their tools and ran away as hard as they could. Nothing ever put out Kanny, so she only began to wonder if they had left anything behind for her to drink, for the day was hot, and in spite of the fruit, she was thirsty. Glancing round, her eye fell on the

KANNY FRIGHTENS THE CARPENTERS

pewter pots, which the men had just filled with beer—for it was their hour for leaving off work—and without hesitation she took a sip. The taste she thought was good. Yes—after another sip—it was certainly refreshing; so from one pot she went to another, until she had emptied them all.

All this time the men were cowering under an outhouse, far too much frightened to interfere with the kangaroo. And even when she was called off, and taken back to her own outhouse, they did their work in fear and trembling for the rest of the day, lest this terrible stranger should come back again.

But as winter came on, poor Kanny's games got fewer and fewer. She had attacks of shivering, which generally ended in fainting fits, and between them she would lie on her bed, looking up sadly at her anxious nurses, who sat by her, stroking her head. At length the weather got so cold that they could not keep her warm in the outhouse, so she was carried in and laid on a soft rug before the kitchen fire. She knew they meant to be kind to her, and though she had hardly strength for the move, she tried to raise her head, and rub it against their hands. But the bitter frost had touched her lungs, and she fell back gasping, and in a few minutes was dead.

The children wept bitterly for their beloved playfellow, whom they themselves buried under a tree; and though time passed and they had other pets, no one ever took in their hearts the place of Kanny.

COLLIES, OR SHEEP DOGS

Shepherds' dogs, when in regular work, are serious animals, and far too busy in their daily life to have time or taste for play. They do not make friends very easily, because they and their masters are accustomed to live alone on the wild hills or great moors, and the sight of other men is strange to them. But they are as useful and necessary to the shepherds, their masters, as any other race of dogs trained to business habits; indeed, the work of keeping a flock together would be quite impossible without them.

The shepherd's dog (or 'collie' as he is called in Scotland) is a beautifully shaped animal, either bright yellow, or black and white, with a curly tail. He is a very quick runner, and a splendid jumper, as he has need to be, when his duty is to follow the sheep into all sorts of rough places, where no man could ever keep his footing. He is regularly sent to school before going out to service, and carefully taught his work, which, in general, he learns very easily; and besides the training he gets in this way, his life soon teaches him to bear hunger and thirst and to do without much food, which is often, in severe winters, very hard to get in distant spots.

As for weeks, and even months, the dog is frequently the shepherd's only companion, the two seem almost to understand the thoughts that are passing through each other's minds without need of speech. One bitter winter's day, about a hundred years ago, a young man was herding his father's sheep on a Cumberland moor, when he fell and broke his leg. Dusk was coming on, the road was lonely, and home was three miles away. To spend the night on the bare heath was certain death; how to get help he knew not. Suddenly an idea came to him: he tied one of his thick gloves round the dog's neck and told him to go home. The dog bounded off,

THE FAITHFUL MESSENGER

and was soon heard scratching at the farmhouse door. At the sight of the glove the farmer at once understood that some accident had happened, and wrapping himself in his plaid, called to his men and prepared to set out. The dog ran first, and after stopping many times to make sure he was not going too fast for the others, led them to where the young man was lying, faint with pain and half dead with cold. A few hours more and it would have been too late to save him.

Sirrah, the favourite collie of Hogg, the 'Ettrick Shepherd,' was, like many people who live in lonely places, rude and unsociable. If a friend patted him, he growled; if anyone admired him, he simply walked away. But, says his master, in spite of these manners, 'he was the best dog I ever saw.' Very little is known of his early history, but when he is first heard of he belonged to a boy down on the border, and was sold by him to a drover for three shillings. The drover brought him northwards, and gave him very little food on the way, so that when Hogg first met him he was very thin, and looking as cross as hungry people often do. At this time he was nearly a year old, with a very dark coat. Partly out of pity, and partly because he thought that the dog looked as if something might be made of him, Hogg offered the man a guinea, which was eagerly accepted, and took his new bargain home. The next day the Ettrick Shepherd began to teach Sirrah his duties, which were evidently quite new to him; but it was wonderful what pains he took to learn, and how grateful he was to his new master. 'He would try every way deliberately till he found out what I wanted him to do, and when once I made him understand a direction, he never forgot or mistook it again.' And besides his care in following out directions, he was wonderfully clever at inventing ways of overcoming obstacles or getting out of difficulties.

One dark night, about seven hundred lambs, which had just been taken away from their mothers, formed themselves into three divisions and rushed away to try to find their way home again. Hogg, and a boy who was with him, did all they could to stop them, but it was no use, for the darkness was so dense you could not see the length of your hand. 'Sirrah!' cried the poor man in despair, 'they're awa';' and so they were, beyond the power of his catching. But Sirrah was cleverer than his master in the matter of catching sheep, and off he started, while Hogg and his helper passed the night in seeking for traces of the lambs, which could not be found, go where they would. At last, when the sun rose, they gave up the chase and returned to the farmer who owned the flock, to tell him of the loss of his sheep, a thing which had never occurred to Hogg before, all the years of his life as a shepherd, neither had he ever heard of it happening to anyone else. On their way back from this unpleasant errand they had to pass a deep hollow or 'cleuch,' as it is called in Scotland, and there, safe at the bottom, were the whole flock of lambs, with Sirrah standing over them. Hogg could not believe his eyes, and at first thought it must be only one of the divisions of the lambs —though even for that he was grateful enough; but when he came to count them there was not one missing. How Sirrah had managed to collect them nobody knew, and of course nobody ever *did* know!

When Sirrah died, he left a son called Hector, to take up his duties. Hector, though not nearly so clever as his father, was a more lively companion; full of whims and freaks, but much attached to his master.

One day in August, Hogg was sent by his master to a farm at the head of the river Ettrick, to bring back some black-faced lambs, intended for next morning's market. Hogg set out, accompanied, of course, by Hector. For some reason or other the lambs were not brought down from the hills till quite late, and the shepherd did not feel at all comfortable at the thought that he would have to drive them the greater part of the way in utter darkness. What was worse, he knew that the lambs, which had only just been parted from their mothers, would be very unruly. However, there was no help for it, the start must be made, and though everything turned out exactly as he had imagined, with the aid of Hector all the lambs were at last safely housed in the fold, and both man and dog were nearly worn out.

As soon as the door of the fold was shut, Hogg went in to his own supper, and then put down Hector's; but the dog was nowhere to be seen. Hogg whistled and called for some time, but to no purpose, and finally he gave it up and went to bed, wondering how in the world he was to drive his lambs to market without the help of his collie.

His first question when he woke was whether Hector had come home; but no one had seen or heard anything of him. What was to be done? Each person suggested something different, till it was decided that the shepherd's father should feed the lambs and get them ready for their walk (shepherds take a great deal of pride in having a smart flock), while Hogg rode as fast as he could back to the farm to ask if Hector had returned there. So father and son left the house together, to bring the lambs out of the fold, and when they reached the door there was poor Hector sitting before it, never taking his eyes off his charges, for fear lest they should run away! There he had sat all the night long in the pouring rain, hungry and tired, a martyr to what he considered his duty, though a wiser dog would have known that, once they were in the fold, he need not trouble about the lambs any further.

The Hoggs had a cat which Hector hated with a deadly hatred, though he was too good-natured to hurt her, however provoking she might be. His way of revenging himself for puss's impertinence towards him was to 'point' her, as if she was a bird, whenever they were in the room together. If annoyed at being watched in this manner, the cat got up and sat in another place, Hector was sure to follow her and begin again; and this went on till he had to go to his work or else fell asleep.

Hector had a very small appetite, and often the only way to make him eat was to bring in the cat and set her to the plate. Then he got furious at her attempting to take what belonged to him; his eyes glared, and his tail stood up straight with anger. When her nose touched the food he could bear it no longer, and began to lap madly, though he never failed to keep to his own side of the dish, and let her pick up anything she could get.

Another of Hector's tricks, and one of which he could never be cured, was to tear round the room like a wild thing, a few seconds before old Hogg finished his daily family prayer; and this was all the more strange as the old man prayed out of his own head, and the length of the prayer varied. It never seems to have occurred to any of the family to shut the dog up, and they all puzzled in vain over the reason for his conduct, when suddenly the true explanation darted into the head of the shepherd himself. 'Hector is all day long pointing the cat; now, when he sees us kneeling with our heads on our hands, exactly as he does, he takes for granted that we are pointing her too, and the moment that he can tell from our father's voice that the prayer is coming to an end, he springs to his feet, saying to himself: "It is no good for them to try, I am the first in the field."'

Like many Scotch dogs, Hector was fond of going to church, when his friends would have preferred him to remain at home, and he also enjoyed taking part in the music. The church at Ettrick, where the Hoggs lived, was small, and the singing very bad, which vexed the shepherd, who determined to try to improve it. This he would very likely have managed to do if it had not been for Hector, who, however carefully he had been shut up at home, always contrived to escape, and would be seen stalking up the aisle to the terror of his master, who knew what he had to expect. The moment the first notes sounded, in struck Hector; higher and higher rose the man's voice, louder and louder became the dog's, while the rest of the congregation hid their faces in their plaids, and laughed till they nearly fell off their seats. For some time Hogg stuck to it from sheer obstinacy, but at length Hector proved too much for him, and he gave up the singing to some one who owned a less musical dog!

Though, as we have seen, Hector was not nearly as intelligent as his father Sirrah, about performing his duties, he would hold tight to what he had been bidden to do, with the stupid obedience of Casabianca. Nothing would turn his attention from his work, or make him lose temper; but he could never learn all sorts of little dodges by which Sirrah managed the sheep, so gave himself twice the trouble that he need have done.

Still, though he was not a practical dog, Hector was very wide awake in many ways, and at any mention of a cat, sheep, or himself, he would cock his ears, and sit bolt upright with the deepest interest. One evening Hogg told his mother he was going over to one of the hills on St. Mary's Loch, to spend a fortnight with a friend, but that he would not take Hector, as he would either disturb them with his singing, or quarrel with the other dogs.

Next morning the river had risen high and made so much extra work, that Hogg was prevented from setting out as soon as he had intended.

When he called Hector to tie him up, the dog was nowhere to be seen.

'Confound that beast!' he exclaimed. 'I'll wager that he has understood what we were saying last night, and has gone to Bowerhope.'

And so he had, though the river Yarrow, which he had to cross, had swollen into such a torrent that it seemed impossible for any dog to swim it. But there he was when Hogg arrived, sitting like a drowned hen at the end of the house, awaiting his master's arrival with impatience.

TWO BIG DOGS AND A LITTLE ONE

Nobody who has ever been the master of a huge, good-natured, silent Newfoundland dog, could bear to have a little fretful, yapping creature as his daily companion, however beautiful it might be. A Newfoundland is large and awkward; he waddles along in a very ungraceful manner, and he will probably never think of moving for visitors, if he takes a fancy to stretch his great body on your doorstep; but he is so strong that the most timid woman would feel quite safe in his care, and so silent that one growl from him rouses the soundest sleeper to a sense of danger. He has webbed feet, and can swim like a duck, and in many places he is almost as good as a life-boat.

Big though he is, a Newfoundland dog is full of life and spirits; full, too, of affection for his master, whom he is always anxious to help and defend. He is easily taught, and untiring in his efforts to carry out his master's wishes, never interfering or quarrelling unless he (or still more, his master) is first attacked, but always on the look out for danger to those whom he loves.

In their own country, Newfoundland dogs play the part that oxen do in Italy, or horses elsewhere. And more; for, wherever they know the road, they can be trusted to draw their carts or sledges piled with wood or hay without being watched by a driver. When they arrive at home they are given their dinner, generally of dried fish, which they much prefer to any other dainties, and then, if necessary, they are ready to undertake the post of night-watchman, or to do anything else that their masters wish.

One weakness, however, Newfoundland dogs have, and that is a love of sheep's blood, which renders it dangerous to keep them in sheep-breeding districts.

A story is told of a man who brought a pure-bred puppy from the north of the island of Newfoundland to his own home, at a place called Harbour Grace. The pup soon became a great favourite with everybody, and especially with the children, and in his leisure moments, when his work was done, was generally to be found in their company. Even the cats rather liked him—he was so big that most likely they didn't think he was a dog at all—he never interfered with anything they did, and was always polite. But the moment he saw a sheep he became another creature. He would chase it until he ran it down, and would even drive it over the cliffs into the sea, and jump in after it! That is, he would jump in if he did not consider the leap too dangerous, for Newfoundland dogs are very cautious. If he did, he would scramble round by an easier road, and reach his prey some other way.

But the puppy, young as he was, was so cunning, that it was often a matter of difficulty to detect his crimes, and so good and useful in other respects, that his master had often not the heart to punish him for them. Besides, as the man felt, it was the nature of the creature, and no amount of punishment would ever alter *that*. The dog must either go, or the man must, as far as he could, keep the sheep out of his way, and when he could *not*, suffer in silence! Still, let him be as careful as he might, the sheep and the dog could not always be kept separate, and then something dreadful always happened. For some time the master thought, however, that Fowler was really cured of his bad habits, for he would pass three young sheep that had been bought without taking the slightest notice of them, never even casting a glance in their direction as he went by, or giving a single lash to his tail. All went smoothly till unluckily one night the servant, whose business it was to shut up the sheep in their shed, and the dog in his own premises, forgot all about it, and the next morning when the sun rose, three dead bodies were found stretched in the yard, with just one little wound in their throats, where the blood had been sucked. The murderer meanwhile was doing his daily work, as cheerfully and patiently as if he had nothing on his mind at all.

Now and then, however, Fowler showed that he had a conscience. Some geese, which had once been wild, lived on the farm, and one of them was very fond of his master, and would even go out to walk with him about the place, strutting side by side with Fowler. They seemed to be the best of friends, and nobody gave a thought to the fact that, next to sheep and dried fish, Newfoundlands love poultry. But they were soon reminded of it. As in the case of the sheep, the geese were one night left at large, by accident. Next morning neither goose nor dog appeared to go their rounds with their master, and search was made. A few white feathers were found in a field outside the grounds, and after some time Fowler was discovered behind a woodstack in the yard, looking, for once, very much ashamed of himself. His master called the dog to follow him, and led him to the field, where the scattered feathers were pointed out to him. Fowler looked once at the man who had detected his crime, then, with a howl of regret rushed away from the field, and no persuasions would get him near his master for many days after.

Of course, most of the stories about Newfoundland dogs have to do with rescue from drowning, for these animals, from their great size and skill in swimming, are more useful in saving life than any man. But here are some very clever tricks of a dog that belonged, a great many years ago, to a Mr. McIntyre in Edinburgh, that have nothing to do either with water or sheep.

This Newfoundland, whose name was Dandie, could pick out his own master's hat from any number of others, and his knife from a heap on the floor. He could even, we are told, detect among a pack, thrown carelessly down, the card chosen by his master. On one occasion he picked up a shilling that had been accidentally dropped by a gentleman present, and concealed it in his mouth, sitting quietly in a corner all the time, and paying no attention to what was going on. At last, when the whole room had been searched, his master said, 'Dandie, find me that shilling, and I will give you a biscuit,' and Dandie jumped straight upon the table, and laid the shilling in front of the owner. Like the dog in the 'Arabian Nights' (only that dog really was a man), Dandie could go out and do his own shopping. His friends, who were many, used to allow him a penny a day, and he took the money regularly to a baker's shop, and bought bread for himself. One day the penny was forgotten by one of these gentlemen, and when Dandie went up to him in the street, he was obliged to confess it. 'But come to me when I go back,' he said, 'and you shall have it.' Some hours after, he heard a great noise at his door, and sent to see what was the matter. It was Dandie, come for his penny. In order to find out what the dog would do, the gentleman gave him a bad one. This the 'Arabian Nights' dog would have found out at once; but people had always behaved well to Dandie, and he was too polite to suspect anything wrong. He went off with his penny to the baker, who refused, of course, to give him the bread. Upon this, Dandie returned to the house he had come from, knocked at the door again, and, when it was opened, laid the penny at the gentleman's feet, with a look that told of the contempt that was passing in his mind. From that day he never took the slightest notice of the man who had made fun of him.

But, after all, it is not only big animals that are of use in the world, as the lion found out when the net that held him was gnawed through by a mouse. Little dogs can be very brave, and very clever, too, as the following tale will show.

No one would ever imagine in looking at the small, short-legged King Charles spaniel that he would be the dog to prevent murder. His long silky ears, which are generally either black, or red and white, hang down on each side of a round, soft little face, very pretty in a lady's drawing-room, but not giving the idea of much intelligence. King Charles II., to be sure, was very fond of this breed, and seldom went out without eight or ten of them hovering about his heels in the Mall; but then he was a person who set great store by beauty, and was apt to value things and people accordingly.

However, here is a true story, in which the tiny King Charles was quite as clever as the best Newfoundland or collie that ever lived.

About the beginning of this century a lady named Mrs. Osburn was occupying a large lonely house in a country place a few miles from London. One day she drove into town to receive a large sum of money which Parliament had voted to her for the discovery of a medicine which was expected to be very useful, and instead of putting it into the bank, as a wise woman would have done, she brought it back in her carriage to her own house. The long day in town had tired her a good deal, and she soon made up her mind to go early to bed and sleep off her fatigues. She was just stepping into bed when a little King Charles, who always slept in her room, became greatly excited, and when she lay down tugged hard at the bed-clothes, nearly pulling them off her in its struggles. She told him several times to lie down, but he paid no attention, only pulling and dragging the harder. At length, finding it impossible to rouse her in any other way, he jumped on to the bed itself, took the clothes in his teeth, and drew them carefully backwards. Then Mrs. Osburn, sleepy though she was, began to think there must be some reason for the dog's very odd behaviour, as he was generally remarkably quiet in his ways. So she got up, put on her petticoat, and took out of a cupboard a pair of pistols, which she always kept ready loaded and knew how to use. She then left her room and went downstairs to see if anything was the matter. No sooner had she reached the floor below than she saw her coachman, fully dressed and holding a candle, coming down the servants' staircase. Without stopping to ask him any questions, she raised her pistol, and informed him that unless he went back up the stairs that moment she would fire. The coachman, who had reason to know that his mistress always meant what she said, and who was, besides, frightened at the discovery of his intentions, obeyed at once, and the lady, feeling that sleep was impossible that night, sat down in a room close by to think what she had better do next. Suddenly she heard a sound of low voices coming towards her, and pushing up the window leaned out and fired her pistol in the direction of the noise. Dead silence followed, and, after waiting and listening some time, she heard no more. Then she made a tour of all the lower rooms, and finding everything secure there, went back to her own, taking the King Charles with her and locking the door behind her. In the morning, as soon as it was light, she got up and went into the garden to the place from which the voices had come ; there she discovered drops of blood, and followed their track till they were lost at a wall at the other end of the garden. She then ordered her carriage—we are not told if it was the same coachman who drove her—and taking the money with her, this time carried it safely to the bank. She then called on Sir John Fielding and asked his advice on the matter. He advised her to dismiss the coachman at once, and to leave the affair in his hands to have it thoroughly inquired into. But nothing more was ever found out; and the only thing that was clearly proved to the satisfaction of everybody was that if it had not been for the little King Charles the lady would have certainly been robbed, and most likely murdered.

CROCODILE STORIES

THE rivers that flow into the Nile are the homes of other dangerous creatures besides hippopotami, and though crocodiles do not attack boats, like their larger neighbours, they are even more to be dreaded by men. They are huge beasts, often twenty feet in length, with great scaly bodies and flat heads, which are furnished with long, terrible teeth.

In proportion to their size they are immensely strong, and even quite a little one has often been known to overpower a man when in the water. He then carries his victim to some favourite haunt and eats him bit by bit.

Now, none of the crocodiles which infest the Nile and its tributaries are bigger or fiercer than those in the district of Gondokoro, where Sir Samuel Baker lay for some time encamped. The natives, who swim like fishes, were constantly in the habit of taking their cattle to pasture across the stream in the morning and bringing them back at night, and it was seldom, indeed, that the passage was made—at the risk of their own lives—without the loss of one of the beasts. Nothing, however, could break them of the habit, not even the fact that two sailors had been carried off in two days, while a soldier, who was working with some other men in shallow water, was seized below his knee. He struggled fiercely, assisted by his friends, and tried to blind the creature; but his leg was so crushed by his enemy's teeth that it was absolutely necessary that it should be cut off.

It was really quite dangerous to go near the river at all, for you never knew when a crocodile might be lurking near. One day several sailors went down to the bank to gather the leaves of a pretty, pink, floating convolvulus, which, when chopped up, made a very good imitation of spinach, and was much relished for dinner. The roots were fast in the mud, but the leaves spread about like water-lilies, and had to be drawn in by their stems. One of the sailors was reaching out as far as he could stretch after a particularly fine young leaf, when a crocodile darted out and seized him by the elbow. The pain was frightful, and the man would at once have fallen helplessly into the river had not his comrades instantly flung their arms round his waist and held him back. Then began a fearful tug-of-war. Neither party would let go, till at last the elbow joint itself gave way. The crocodile went off triumphantly with the hand and forearm, and the sailor was carried off to be doctored in the camp.

This was bad enough, but sometimes worse happened. It was no uncommon thing, if one person was out alone on any errand that took him near the river, for nothing more ever to be heard of him. If a woman was washing at the bank in the shallow water, her legs might be seized, and she would be dragged underneath before anyone in the boat moored close by knew what had occurred. This once actually happened to a negro girl; and how she had met her end was afterwards proved in a ghastly manner.

Life was made such a terror by the constant and often unseen presence of these crocodiles, that Baker lost no opportunity of killing all he could with a small rifle, called 'the Dutchman,' kept solely for this purpose. It was so very accurate in its aim, that at a hundred yards it was possible to hit a crocodile which was lying on a sandbank in the two places where death was immediate, behind the eye, or through the shoulder; often the creature never even stirred, but lay dead in its place.

Baker had one day been out on some business, and

was riding back to his own quarters further up the river, when he saw a large crocodile lying out in the stream, with its head above water. In order not to be observed before he could get near enough to aim, Baker dismounted, and crept softly away from the bank, which he then struck a little lower down, where a clump of rushes would conceal him from view. Almost crawling along the ground, he reached the spot, about four feet above the river, and took careful aim behind the crocodile's eye. The animal gave a start, and turned over on its back, where it lay without moving, with its legs above the water, which there was only two feet deep. Baker, of course, thought it was dead, and taking the rope which he always carried on his horse, told two of his men to go into the water and tie it up securely. While this was being done, a third man was sent off on horseback to the camp to bring back help, for long experience had taught them that, though a crocodile may really be shot through the brain, the muscular movements, both of legs and tail, will gradually cause it to slide from the bank back into deep water.

The men did as they were bid without shrinking, for they, too, had seen the fatal shot, when suddenly the scaly tail began to move. Trembling with fear, they cried out that the animal was still alive; but Baker told them it was all nonsense, and bade them be quick and finish what they were at. The men being on the spot, however, knew much better than their master on the bank, and the crocodile's struggles soon got so strong that they could hardly hold it. All at once it gave a great yawn, and, had it not been for dread of punishment, they would have dropped the rope in a fright and left the animal to its fate. Another bullet in the shoulder checked its struggles, and by this time the men galloped back with more ropes. Even now its strength was by no means exhausted, and it did not submit easily to its fate; but at last it was safely landed on the bank, and a sharp blow with an axe divided its spine.

When the crocodile was found to be dead without a doubt, its stomach was opened. Among other things, too horrid to mention, there were found, inside, two armlets and a necklace that had probably belonged to the

FINDING THE NECKLACE

negro girl who had disappeared so completely while washing in the river.

Further south still, beyond the great lakes, is the River Zambesi, whose branches swarm with alligators, a kind of crocodile, and quite as dangerous.

Fifty years ago, when Livingstone was travelling up one of these rivers, the Leeambye, which falls into the

Zambesi, he came to a whole district where the children were constantly being snapped up by these frightful creatures, when they went to play on the edge of the stream. A blow from the tail of an alligator would knock down a child or a calf that had come to drink, and then the great flat head would be thrust out of the water, and the victim was pulled in without any chance of escape. One day, a man in Livingstone's caravan was swimming across one of these rivers, when an alligator caught hold of his thigh, and dragged him below, but not before he had managed to get out a knife he carried with him; and as he sank he stabbed the alligator in the shoulder. Smarting with the pain, the alligator loosened his hold, and the man came up to the surface, not very much the worse, but with marks on his thigh that he never got rid of. Luckily for him, his tribe had no superstitions about bitten people; but in some of the other places visited by Livingstone, any man who has received a bite from an alligator, or has been splashed by his tail, is considered unclean, and chased out from his fellows. They think that merely to look at the wound would cause a disease of the eyes. If the bite happens to be caused by a zebra, the sufferer is not only obliged to fly himself, but to take his wife and family into the desert. The Barotse tribe have no objection to eating alligators, which most people would find very 'strong' meat; and Livingstone tells that one of them complained to him of an alligator carrying below a wounded antelope which had taken to the water. 'I called to it to let my meat alone,' said Mashuana, 'but it would not listen.' So, in revenge, Mashuana speared another alligator, and ate it himself.

LION-HUNTING AND LIONS

IN the country of the Shoolis. which is one of the districts drained by the rivers that flow into the Nile, hunting is carried on under very strict rules. In most savage places men go and kill what beasts they like when they are hungry, but among the Shoolis this was not allowed, and everything was arranged by a grand council of the villagers, presided over by the chief.

Sometimes, when all was settled, the chief would give a party before the hunt, and as many as a thousand guests would arrive from the villages round, clad in their smartest ostrich feathers and best leopard-skin cloaks. Then they would dine off freshly killed oxen, and afterwards a sorcerer would work them spells, first to preserve them from accidents, and then to bring them plenty of game.

So, when Baker's people began to want fresh meat, he arranged with the chiefs of the tribe for a hunt, and this was how they set about it.

On the day appointed some thousands of people—men, women, and even babies—assembled at the place of meeting, each man carrying a net twelve yards long and eleven feet high, the boys bearing lances, suited to their sizes.

They marched several miles, and as they went along other natives would silently join in, till the company reached a wide treeless grass country, broken up by many streams. Here the nets were set up in a line about a mile and a half across, and every man went to his station on each side of the net, hidden by the long grass, tied together at the top. By the rules of the chase all the beasts killed before each twelve yards of net belonged to the owner of the netting, who had to pay the tribute of a hind leg from every animal to the man on whose ground the hunt happened to be.

When all was ready a whistle, taken up and repeated for two miles down the line, gave the signal. The men touched with their fire sticks the dry grass, and soon little columns of smoke were seen rising into the air. Not a native was in sight, and Baker, who was standing beside a tall ant-hill, concealed himself as well as he could.

A fresh breeze was blowing, and the fire spread rapidly with a loud roar, and the Englishmen began to look to their guns. A huge rhinoceros made its appearance first, but turned off to the right, and no more was seen of him. After that the rush became thick and fast: leopards, antelopes, hartebeests, dashed wildly along, followed closely by a lion and lioness, far too frightened themselves to think of attacking the antelopes, who, on their part, gave no heed to them. Baker aimed at the head of the lion, but before he could shoot a woolly black head bobbed up between him and his prey. He had forgotten the natives lying in the grass near the nets, and the lion swept by and bounded over the stream, and no more was heard of him!

Bad though the fire was for the animals, things were not much better for the Englishmen, who were nearly blinded by the smoke, and fired wildly in the hopes of killing something. At length the flames reached the shore and at once died down, and when the smoke had a little blown away they all came out from their hiding places to count the spoil. Antelopes had suffered the most, and enough of them had been killed to supply the people for many days. Buffaloes had been seen, but they had headed in another direction and escaped; while, as for a rhinoceros, the net had yet to be made that would stand up before his great carcass. Some days later, when Baker's share of the hunt had all been eaten by his men, he got leave to go with a few natives and shoot what was necessary to supply his camp. But the animals were still so wild after the fire that it was impossible to get near enough for a shot, and at last the owners of the land proposed that they should fire the grass to windward, as before.

This time Baker took care to choose a position with some swampy ground in front, so that when the fire reached him it would be stopped, and he would no longer have the smoke blown into his eyes.

Again, antelopes were more numerous than anything else, but none of them came within reach of Baker's own gun. After waiting a little, however, he saw a fine specimen moving quietly towards him down a bank into a dip, and made ready for a shot. The antelope was just in the act of jumping down from the slope into the hollow when he almost tumbled against a huge lion, which had come up from the other side, and was flying before the fire. Both lion and antelope were so much startled by the shock that they bounded away in opposite directions, the lion taking a line through the tall grass, which would bring him straight in front of Baker.

For a few minutes all was silent, Baker leaning against the ant-hill, with one gun in readiness and another by his side, and the two black boys crouching on the ground at his feet.

Suddenly a rustling was heard in the grass, and all three waited breathlessly till the head of a lioness appeared, coming slowly but steadily towards the spot where the two boys were sitting.

A ball in her chest stopped her proceedings for a moment, and she rolled over three times, uttering terrific roars all the while. Then she got up, apparently none the worse, and, lashed to fury by a second shot, advanced by high leaps towards the frightened boys.

On this, Baker, who had till now been hidden behind the ant-hills, snatched up his spare gun and stood in front of his cover. The lioness was startled by this movement, and half turned, receiving as she did so a charge of shot in her hind quarters. This decided her to retreat, and the grass soon hid her from sight, though they still heard her groaning.

Then some of the other men came up, and were hastily placed in line to receive the lioness when she should make her charge.

A shot soon brought her out, charging in those tremendous leaps so frightening to see, and the spears thrown by the natives missed her entirely. There was nothing for it but flight, and in a moment the black men were tearing for their lives in every direction. But a shot from Baker's breech-loader right in the chest rolled her over a second time, when she had almost reached him, and a ball at the back of her neck, fired at twelve yards distance, at last put an end to her struggles.

Inside her stomach was found a freshly eaten antelope, which the black men, who were not particular, begged to have for their dinner. After this it is not surprising to hear that they were prepared to eat the lioness herself, while the white men took the other antelope for their share.

Nearly sixty years have passed since Dr. Livingstone sailed for Algoa Bay, whence he was to start for his missionary travels into the centre of Africa. His journeys were made either by ox-back, or on foot, and at first the natives despised him for his size, which was much less than theirs; but it was not long before they learnt to take a different view of the white man who had come among them.

In the middle of the Bechuana country, which is bordered on the west by the great Kalahari desert, lies a village called Mabotsa. When Livingstone first visited Mabotsa, in the year 1843, he found all the people living in terror of the lions, which would not only invade the cattle-pens by night but attack the herds by day. This happens so seldom, unless the lions are very hungry, that the villagers explained it to themselves as being the result of a spell wrought by a neighbouring tribe, which had given them into the power of the lions; and as they stood rather in dread of the fierce beasts, the lions had it all their own way.

Livingstone, however, had other ideas about the matter. He knew that if they could only manage to kill *one* lion, the rest would go and look for their dinner elsewhere. So, when the herds were again attacked in the grazing ground in broad daylight, he persuaded a large body of men to come out with him and punish the robbers.

It was not long before the lions were seen comfortably seated on some rocks which jutted out from the plain, thickly covered with trees. Livingstone ordered his men to surround the hill completely, at a distance, and then gradually to approach nearer and nearer, so as to make a close circle. A shot from the native schoolmaster hit one of the animals lying on a spur of the rock, and very much surprised him. He bit fiercely at the place, as if he thought he had been stung by an insect, then got up, and clearing the circle at a bound, vanished into the distance.

Two other lions followed his example, and got off without a scratch, for neither Livingstone nor the schoolmaster could fire from below, without risk to the men above; and the fear of magic seemed to have so paralysed the natives that they never even thought of using their spears, as was the custom of their nation. Seeing it was hopeless to get the men to act, Livingstone called them off, and gave the word to return to the village. On their way past the foot of the hill they came suddenly upon one of the animals sitting on a rock behind a small bush, and this time Livingstone resolved to take the law into his own hands, and not to trust to the Bechuanas. When thirty yards distant, he fired both barrels at the sitting lion, straight through the bush, and heard a cry of triumph from the natives. The lion might be shot, but he certainly was not dead, for his tail stood up in a threatening way. Thinking that the men would run forward before it was safe, and would attract the notice of the wounded beast, Livingstone called out to them to stop till he could reload his gun. He was just putting in the charge, when a cry caused him to look round. The lion was in the air, close to him. In sweeping by he seized the missionary's shoulder between his teeth, and brought him to the ground from the hummock on which he was standing, growling and shaking him all the while. After the first moment, Livingstone felt nothing. He lay still as if in a dream, not even frightened as to what was coming next, and quite unconscious of any pain; but gazing at the lion who stood above him, keeping his great paw on the head of his prey. From the position of the lion's eyes, Livingstone at once guessed that he was watching the schoolmaster Mebalwe, who, braver than his fellows, was trying to shoot him with a clumsy old gun, from a distance of fifteen yards. Both barrels missed, and with a roar the lion let go Livingstone and leapt upon Mebalwe, whom he caught by the thigh. On this a man thrust a spear into the lion from behind, and was instantly seized by the shoulder; but the bullets which were poured into the brute from other quarters now took effect, and the great beast fell back stone dead.

In the excitement of the battle, which, after all, had lasted only a few moments, the wounded men had hardly been aware of pain; but when they were all satisfied that their enemy was really dead, they began to examine their injuries. Livingstone had been partly saved by his jacket, which had received most of the poison of the

THE LION WAS IN THE AIR CLOSE TO HIM

lion's teeth, so that, although it was bad enough to have the bones of his arm crushed into splinters, the eleven flesh wounds on his shoulder, healed without leaving any ill-effects. The other two men, on the contrary, who had had nothing to protect them, suffered to the end of their lives from strange pains in the wounded parts, which were always particularly violent at the season of the year in which the lion had bitten them.

The next morning a great bonfire was made in the village, and the lion solemnly burnt; and from that moment the spell was pronounced broken—and the lions went away.

The idea that lions are the bravest of all animals dates from the time when very little was really known about them. Anyone who reads Mr. Livingstone's travels in South Africa will find that he tells a widely different tale. According to him, no *single* lion will ever attack a man by daylight or even moonlight, unless he is first attacked himself, or almost starving. Even on a dark, rainy night, the dread of falling into a trap is enough to keep him from assaulting any animal tied to a tree, and therefore at his mercy. It is curious how fear of pitfalls never leaves him! One day, an Englishman's horse, which had bolted and thrown its rider, was caught by its bridle in the fork of a tree and held fast. For two days search was made for it, but in vain. On the third they came upon the missing creature by accident, quite safe and sound, though all round it were the marks of lions' paws! Any animal tied up seems to act as a charm against lions, by night as well as by day—they will not even attack a sheep, lest something unknown and terrible should be the consequence.

As a rule, unless they have little ones, nobody need be afraid of lions from sunrise to sunset! Livingstone and his family used often to meet them in their walks outside the camp, and after staring with surprise for a few seconds, the lion would turn slowly round and cautiously move away, keeping his head turned over his shoulder. When he had got to a little distance he would break into a trot, and finally, when he thought no one saw him, and he had no character to keep up, he bounded away as fast as he could. Indeed, the missionary positively declares that a man runs a much greater risk in crossing a London street than he is ever likely to do from the king of beasts—unless, of course, he is being hunted.

Besides, no young lion will look at a man as long as he can get any other food. It is only when he is old and loses his teeth that he gives up hunting wild game, and, driven by hunger, ventures down to the villages to catch goats, mice, or any stray man that may happen to be about. The village people know this so well, that when goats are found missing from the herd they will shake their heads, and say to each other, 'His teeth are worn, he will soon kill men,' and set about arranging a hunt immediately.

Lions generally attack their prey by leaping on to its flank from behind, though they will sometimes fly at the throat. A friend of Livingstone's tells a story of a sight he saw on the banks of Limpopo river, when on a hunting expedition in the year 1846. He was riding along with another man in search of game, when a fine water-buck jumped up from the reeds in front. The Englishman dismounted, in order to follow it, and by doing so disturbed three large buffaloes, which stood and looked at the strange white thing they had never seen before. A ball in the shoulder of one awoke them from their stupor, and they galloped away, closely pursued by the hunters. Suddenly three huge lions sprang on the back of the wounded buffalo and dragged him to the ground. The two Englishmen crept softly up till they were within thirty yards of the group, when they knelt down and fired from their single-barrelled rifles. One lion turned, seized a small bush between his teeth, and fell dead right on top of the buffalo; another bounded off as fast as he could, and the third took no notice whatever. This gave the men time to reload, and a ball passed straight through his shoulder blade. Then he thought he had better retire; but he had not gone very far before a bullet in his heart put an end to him.

Lions are fond of hunting in families, sometimes six or eight at once, and in any country where game is abundant lions may also be looked for. They will very rarely molest a full-grown animal, if they can get hold of a young one, and if a buffalo mother finds a lion trying to carry off her calf a fearful fight takes place, in which, if the lion is alone, he is pretty certain to get the worst of it. 'One toss from a buffalo bull,' says Livingstone, 'would kill the strongest lion that ever breathed.' Even a number of lions have been known to be kept at bay by an equal number of buffaloes, who put the little ones and mothers carefully in the rear, and stood with their horns steadily turned to the enemy.

But, as old Topsel says, 'There is no creature that loveth her young ones better than the lioness, for both shepherds and hunters, frequenting the mountains, do oftentimes see how irefully she fighteth in their defence, receiving the wounds of many darts, and the strokes of many stones, standing invincible, never yielding till death; yea, death itself were nothing to her, so that her young ones might never be taken out of her den. It is also reported, that the male will lead abroad the young ones, but it is not likely that the lion, which refuseth to accompany his female in hunting, will so much abase his noble spirit as to undergo the lioness' duty in leading abroad her young ones. In a mountain of Thracia,' he goes on to relate, 'there was a lioness which had whelps in her den, the which den was observed by a bear, the which bear on a day finding the den unfortified both by the absence of the lion and the lioness, entered into the same and slew the lion's whelps, afterward went away, and fearing a revenge, for her better security against the lion's rage, climbed up into a tree, and there sat as in a sure castle of defence. At length the lion and the lioness returned both home, and finding their little ones dead in their own blood, according to natural affection fell both exceeding sorrowful, to see them so slaughtered whom they both loved. But smelling out by the foot the murderer, followed with rage up and down until they came to the tree whereinto the bear was ascended, and seeing her, looked both of them gastly upon her, oftentimes assaying to get into the tree, but all in vain, for nature which adorned them with singular strength and nimbleness, yet had not endued them with power of climbing, so that the tree hindered their revenge, gave unto them further occasion of mourning, and unto the bear to rejoice at his own cruelty, and deride their sorrow.

'Then,' continues Topsel, who writes in very long sentences, 'the male forsook the female, leaving her to watch the tree, and he like a mournful father for the loss of his children, wandered up and down the mountain, making great moan and sorrow, till at the last he saw a carpenter hewing wood, who, seeing the lion coming towards him, let his axe fall for fear. But the lion came very lovingly towards him, fawning quietly upon his breast with his forefeet, and licking his face with his tongue; which gentleness of the lion the man perceiving, he was much astonished, and being more and more embraced and fawned on by the lion, he followed him, leaving his axe behind him which he had let fall, which the lion perceiving went back, and made signs with his foot to the carpenter that he should take it up. But the lion perceiving that the man did not understand his signs, he brought it himself in his mouth, and delivered it unto him, and so led him into the cave where the young whelps lay all imbrued in their own blood, and then led him where the lioness did watch the bear. She, therefore, seeing them both coming, as one that knew her husband's purpose, did signify unto the man that he should consider

THE WOODMAN AND THE LIONS GET THE BEST OF THE BEAR

of the miserable slaughter of her young whelps, and showing him by signs that he should look up into the tree where the bear was, which when the man saw, he conjectured that the bear had done some grievous injury unto them. He therefore took his axe, and hewed down the tree by the roots, which being so cut, the bear tumbled down headlong, which the two furious beasts seeing, they tore her all to pieces. And afterwards the lion conducted the man unto the place and work where he first met him, and there left him, without doing the least violence or harm unto him.'

Topsel, and the ancient authors from whom he quotes, who only knew lions by hearsay, had a much higher opinion of the tribe than Livingstone and modern travellers, who have made their personal acquaintance. He says nothing of their dread of man or ever-present dread of pitfalls! To Topsel, the lion is just a mass of noble qualities, and an example to all men in the matter of family affection. But, then, people often seem different, to those who know them best, from what they do to strangers!

'Neither do the old lions love their young ones in vain and without recompense,' he ends up, as the moral of the last story, 'for in their old age they requite it again; then do the young ones both defend them from the annoyance of enemies and also maintain and feed them by their own labour; for they take them forth to hunting, and when, as their decrepid and withered estate is not able to follow the game, the younger pursueth and taketh it for him; having obtained it, roareth mightily like the voice of some warning piece, to signify unto his elder that he should come on to dinner, and if he delay, he goeth to seek him where he left him, or else carry the prey unto him. At the sight thereof, in gratulation of natural kindness, and also for the joy of good success, the old one first licketh and kisseth the younger, and afterward enjoy the booty in common between them.'

It is not often that a dog which has been carried off by a lion comes back to tell the tale, yet that is what happened to Blucher when he went hunting in South Africa with Mr. Selous in the year 1882.[1]

One night his master was lying awake, reading, in his camp in Mashonaland, when he heard Blucher and the rest of the dogs set up a furious barking lower down the valley. Selous sat up and listened, and as the noise seemed always coming nearer, he called one of his men and asked him what was the matter.

'It must be a lion,' he said; 'Blucher would never retreat like that before a hyena'—for hyenas are great cowards and easily frightened.

As he spoke Selous jumped down from his bed on the waggon, and, followed by Norris, walked to the edge of the camp. It was pitch dark, and they could see nothing; but suddenly the barking ceased, and some large animals came tearing past, while the puppies dashed in for shelter between the men's legs, almost upsetting them. Then one of the natives rushed up from their camp a little further off, shouting 'Lion! lion! Lion has caught big dog.'

Selous felt very sorry for the loss of his old friend, but he took comfort in thinking he must have been killed in a moment, or some yelps of pain would have been heard.

However, the hunter armed himself with his rifle and went back with the native to his camp, where all the men were sitting up, talking softly round big fires. One or two declared they were sure some animal was creeping about in the dead leaves outside the camp; but as, after a search, nothing could be found, Selous got tired of standing about in the cold, and went back to his own waggon. He was just dropping off to sleep when the puppies again set up a furious yapping, and a Kaffir shouted out, 'Here's the

---

[1] *Travel and Adventure in Africa.* By F. C. Selous.

lion, he has taken the skin,' for the skin of a freshly killed antelope, which had been hung up inside the camp, had disappeared altogether. The Kaffir boy, who had been sitting behind the fire, had seen the lion come straight through a hole in the fence close to the dogs, and quite near the horses, and pick up one of the three skins rolled up on the ground. The lion does not seem to have noticed (or smelt) the horses, or they him, which proves that there

THE LION IN THE CAMP

is no truth in the story that horses always scent lions from a great distance.

Notwithstanding all this excitement, Selous, who was very tired, returned, for the third time, to bed, and for a time all was pretty still. Then, again, there was heard the dash of the puppies from outside the camp, and one of the men observed that a lion must be about. On this Selous got up, and looking at the antelope skins discovered that another had been taken away. So he carried off the last remaining one and threw it for safety on the waggon where he himself slept. As the dawn was not now very far off, he lit a candle and took up his book.

Not an hour later he was aroused by a great rattling in the direction of a large packing case outside the camp, where some tools had been left lying. He sprang up, with Norris after him, and in the dim light he saw the white case being shoved about, though it was still too dark for him to make out the lion. However, Selous aimed straight at the case, and absolute quiet followed his shot; but only for a moment, then the case began to move more wildly than ever, till a second shot caused its dancing to cease.

Everybody felt by this time that they never wished to see a lion again, and dogs and men alike stretched themselves out wearily. But it was barely half an hour later when all the noises began afresh, and the waggon itself was shaken. The lion had positively returned to the charge, and not finding any more new antelope skins on the ground had been obliged to put up with an old one, which was hanging to dry on a platform between two poles. When he got on to that platform, which he probably did with a spring, he was within six feet of Norris and another boy.

Except for the sound of the lions crunching the leg bones of the antelopes (which had been left in the skins) in the open ground by the river, nothing further happened that night. With the first streaks of dawn Selous got up and peered about him; in the faint light he made out something which he took to be an ant-heap, but it turned out to be a lion, and nearer the river was the lioness and two or three little dots of cubs.

Thinking that they had gone to drink, and would soon be seen climbing up the steep bank which overhung the stream, Selous crept after them in order to get a better shot. But when he reached the place where they had disappeared no lions were there! In vain he sent for his horse and galloped backwards and forwards down one bank, while Norris did the same on the other. The lions had gone down the river under cover of the high bank, and had got safely away to the forest.

They were all, of course, determined not to spend such another night as the last, so they set a gun trap for stray visitors, and baited it with a large piece of meat. They had just finished their preparations when a cry was heard from one of the Kaffirs, and turning round Selous saw poor Blucher come slowly and painfully up by way of the river. He was covered all over with wounds, and had four holes in the loose skin of his neck, where the lion had seized him. How he had escaped, or why he had waited so long before returning to camp, no one ever knew; but he wagged his tail feebly at the voice of his master. They did everything they could for him, and in time his wounds healed; but he never got really well, and only grew thinner and thinner till one morning he was found quite dead.

ON THE TRAIL OF A MAN-EATER

FIFTY years ago, when Colonel Gordon Cumming, then a young man, was sent out to join his regiment in the country of the Mahrattas, India was full of tigers, bears, wild boars, and other fierce beasts, who were the terror of the native villages. The district is hilly and rocky, and abounds in rivers and thick jungles, which afford shelter for even the largest animals, who would come down at night and carry off goats, oxen, or even men. The English soldiers asked nothing better than to be allowed to put a stop to this state of things, and many were the adventures that happened to them in their shooting expeditions.

Here and there, indeed, an old man was to be found who, like old Kamah, was at peace with the tigers, and looked on any injury done them as an insult to himself. ' I have no quarrel with tigers,' he exclaimed indignantly, when the hunters found him beating his fifteen-year-old son for shooting a tiger who had carried off a tame buffalo. ' I live in the jungle, and the tigers are my friends. I never injured one of them, they never injured me ; and while there was peace between us I went among them without fear. But now, now——'

Kamah's view of the tigers was, however, not common; and, in general, the natives would gladly turn out to help in hunting down their natural enemies.

Sometimes platforms were built in the trees, carefully chosen near the track the tiger was likely to follow, but this was not always very safe, for tigers are great jumpers, and, when maddened by wounds, have been known to pull down a man in a tree. This happened once in an expe-

CUMMING'S CAP FRIGHTENS THE TIGER

dition of Colonel Gordon Cumming's, when he took his stand with his gun-bearer on some branches which grew about eight feet up the stem of a tall tree. The tree was on a slope, close to a small rocky ravine, and beyond the ravine was a jungle in which lay the tiger.

The two men had not waited long before they saw a black and yellow body moving at a brisk trot, on the further side of the ravine, having been roused from his lair by the natives. Gordon Cumming fired, but the shot only wounded the tiger slightly, and he turned and plunged into the jungle. The hunter, having fired the other three balls after him, without touching him at all, gave him up for lost, and did not even reload his gun. Suddenly there was a cry that the tiger was coming back, and, sure enough, there he was crossing the ravine, and making for the slope. When he reached the tree he stood still. There was no time to load; all they could do was to sit quiet, hardly daring to breathe, hoping that the tiger would pass them by. And most likely he would have done so had not the native whispered in a low voice that the tiger was below. The beast looked up, and with a flying leap landed on the trunk of the tree, close to the man's legs. Digging his long claws firmly into the bark, he seized the poor fellow's waist-cloth in his teeth, and dragged him to the ground, biting him severely in the thigh as they rolled over together. The Scotchman did all he could to scare the tiger, by shouting and by flinging his cap straight in his face; this startled the animal, and, letting go his prey, he ran down the hill. Gordon Cumming then came hastily to see if Foorsut was badly hurt, and found that there were twelve severe wounds at the back of the thigh. He was put on a litter of twisted boughs, and carried back to the camp to have his wounds dressed by a native doctor, and then the officers both mounted an elephant and went in chase of the tiger. He had not gone very far, and one shot soon disposed of him. He was a good, large specimen, about ten or eleven feet long, and made a fine skin.

As to the unlucky Foorsut, who had nothing but his own folly to thank for his injuries, he seemed doing well, and to have escaped any injuries to his leg bone. His friends were quite happy about him till the morning of the second day, when his fingers began suddenly to twitch, and by four that afternoon Foorsut was dead.

Some time after this adventure Colonel Gordon Cumming was sent to do some work in the country beyond the Nerbudda, which was, at that period overrun with tigers. The animals found shelter in the broken ground, covered with high grass and sharp, prickly shrubs, from which they would steal out to attack cattle and sometimes men.

One day a villager came to Colonel Gordon Cumming, and told him that a tiger had rushed out and killed a man who had been gathering gum from a tree, in company with two friends; and they, being unarmed, could do nothing to save him.

As it was almost sunset, and the man was known to be dead, nothing was done that night; but as soon as it was light next morning, Gordon Cumming, with two officers, rode off to the jungle. Here some men were waiting with guns and elephants, and the place of attack being arranged, they went first to the gum trees where the man had met his death.

His body was still lying on the ground, bloody where the tiger's teeth had torn it, but otherwise untouched, which looked as if the tigers had all gone elsewhere. However, men were sent up the trees to report if anything was to be seen, and the British officers took up their positions and advanced into the jungle.

They had not gone very far before a huge tiger sprung out of a watercourse where it had been hiding, and dashed up the bank. He was too far off to hit with certainty, and the bullets sent after him only put him out of temper, and he growled loudly as he disappeared into the nearest thicket. The hunters followed on his track at a safe distance, and once they caught sight of him, but again he was off, and was reported by the men in the trees to have hidden in some bushes on the edge of a ravine. Slowly the elephants followed in his trail, when suddenly

THE ELEPHANT TRIED TO GORE THE TIGER WITH HIS TUSKS

the tiger broke away from his hiding place, about eighty yards away. They fired, and this time he was touched, but not badly. Taking refuge again in the bushes, he was lost to view.

Reloading their guns, the officers were entering the jungle, when the tiger started up in front, not twenty yards away, and came on with a rush. A bullet checked his advance for a moment, but he charged again, and the riders expected to see him the next instant grappling with the elephants. But, instead, he sprang right through the animals, and disappeared in the ravine.

Very cautiously the British officers went after him, searching each patch of grass and clump of bushes, lest he should be hidden there. But no tiger was to be found anywhere. At last they had reached the top of the ravine, which was almost filled by a huge green bush, and, though by this time they all felt nearly sure the tiger must have escaped them, they determined to know what was behind that bush. The green leaves were already tickling the elephant's trunk, when it was seized by the tiger, who held it fast between his teeth, while he dug his claws deep into the animal's face. The elephant, mad with pain, gave a frightful shriek, and tried to gore the tiger with his tusks, which was not so easy; and in his frantic plunges the driver ran a great risk of being thrown from his seat, and trampled to death between them. At length a furious shake forced the tiger to loosen his hold, and he turned and fled down the ravine, while the elephant danced with passion on the bank. When he had grown a little calmer, they all turned and went after the tiger.

He was at last brought to bay a hundred yards further down, and this time sprang straight at the head of one of the other elephants. But the shots he had already received were now beginning to tell, and his attack was not so fierce as before. Another ball ended his struggles; he let go the elephant's trunk, and, rolling heavily on the ground, turned over quite dead.

GREYHOUNDS AND THEIR ARAB MASTERS

If we travel about from one country to another, we shall find that each one has a particular kind of dog which is considered useful and precious above all others. In Scotland it is the collie which is most prized, in the high Alps it is the St. Bernard, while in Greenland no one would get on at all without the Eskimo dogs, who draw sledges and do quantities of other needful work, and in Newfoundland there are very few houses which cannot boast of one of the huge black good-natured dogs who are equally ready to be nurses to the children, or to jump into the water to save a drowning man.

Now, in the high plains of Kordofan, which lie to the west of the White Nile, the greyhound or wind dog, as it is called by the Germans, is held in great honour. If you walk through any of the villages, you will see three or four greyhounds lying before the door of every hut, each one more beautiful than the other. They are the village policemen, and guard the people from the fierce leopards and hyænas which steal down at night from the caves where they sleep all day, and prowl round in search of a supper. Like their enemies, the greyhounds sleep during the long hot hours when the sun is up, but the moment he sinks, and the quick darkness of the tropics comes on, they stretch themselves and begin to set about performing their duties. There is no quarrelling or confusion—each dog seems to have his post, and he goes to it at once. If the village is walled in, a certain number will betake themselves to the walls, while others mount to the thatched roofs of the low round huts, and lie quietly down, their eyes open and their ears at full cock, waiting to catch the slightest sound. Sometimes, when the hyænas and leopards have been particularly fierce, a dog or two will take up positions in the outskirts of the village, to give the first warning of danger. Here and there, early in the evening, a bark or a growl may be heard, but as the darkness deepens these die away and all is still, till suddenly the village is awakened by the sound of a battle. Rarely does a night pass without something of the kind. In a few minutes every dog is gathered at the place where the enemy has come up, and directly he is dead on the ground, they leave him there, and go proudly back to their posts. Only on one occasion is their courage known to fail, and that is when the robber turns out to be a lion. Then the conquerors of leopards and hyænas tremble with fear, and shrink howling into some safe corner or hide themselves amongst the thorny hedges that surround the village.

Twice in every week the dogs were given a rare treat. Very early on these mornings the sound of a horn was heard, and then what stir there was among them! From each house three or four came bounding to the place from which the noise proceeded, and in a few minutes after the first blast, fifty or sixty dogs were gathered together. Like eager boys they crowded round the man, jumping up on him or running to and fro with excitement, howling, barking, yelping, snarling, jostling each other to get nearest to the trumpeter, and really behaving as if they had all gone mad. In the midst of all this confusion the young men arrived, bearing in their hands lances and ropes, and sought out their own dogs from the throng. From five to six were led by each man, and hard work it was to get the restless creatures leashed together, jumping and barking all the while with joy! At last all was ready and the hunting procession moved out of the village, and very fine it was! They seldom went far; the neighbouring woods were full of game, and thanks to the skill and quickness of the dogs, the men had an easy time of it. The leash was slipped, and the dogs dashed into the thickets, and soon reappeared, bringing with them all sorts of game—bustards, guinea-hens, or anything else that they happened to come across. If they spied an antelope, five or six would join to chase him, and it was seldom, indeed, that he got away. At the end of the day the spoil was counted over, and was found to consist of antelopes, hares, birds, and often wild animals, such as pariah dogs or desert foxes.

The greyhounds are the pride of the dwellers in the Kordofan desert, and every man thinks his own dog the most beautiful and clever in the world. This breed is not to be found among the Arabs who live among the marshes that border the Nile, and if by any chance one of the desert highlanders wanders that way with his dogs, one or two are sure to be snapped up by the crocodiles. Those dogs who are born and brought up on the banks of the Nile seldom fall a prey to these terrible creatures. If they are thirsty, they never drink till they have looked carefully up and down to make certain that their dreaded enemy is not lurking close at hand. But the desert dog, who knows nothing about rivers or crocodiles, leaps gaily into the stream, and is dragged underneath by his destroyer.

In the west of the Sahara, dogs, as a rule, are only valued for their uses, and are not treated at all kindly; but all the care and affection that the Arab has to give, he bestows on the greyhound. His dog is the apple of his eye, and the two almost eat from the same dish, and share the same sleeping mat. A Sahara Arab will travel joyfully twenty or thirty miles to find a suitable wife for his beloved companion.

A really good greyhound is so swift that it can overtake a gazelle in a very short time; and there is a saying among the Arabs, that if one catches sight of a gazelle grazing, he will catch it before it has time to swallow the food that is in its mouth.

THE SUMMONS TO THE HUNT

The little greyhound pups are petted from the time they are born, and the villagers bring presents of milk and other things to the mother. There is no flattery they will spare, and no promise they do not give, for the chance of getting one of the puppies for their own. 'I am your friend, my brother,' they will say; 'grant me, I pray you, the favour that I ask. When you start for the hunt, I will go with you; I will serve you and help you as a friend may.' Then the master of the greyhound answers that in seven days he will make up his mind whether he will part with the puppy or not, and till then the man must wait. This is because in every litter of greyhound puppies, one is always better than the rest, and in order to find out which is the cleverest the owner will take it away from its mother's side before it is seven days old, and see if it can get back by itself. If it can, he believes the pup will turn out a great prize, worth the best negro slave that could be offered him. It would be dreadful indeed, if he found he had given away such a treasure!

After fourteen days the little fellows are fed upon the milk of goats or camels, with as many dates as they like to eat, and as soon as they are three or four months old, their education begins. The boys let out some small animal under the puppy's nose, and while he is still watching it in a puzzled way, set him on to catch it. It does not take long to awaken his sporting instincts, and in a few weeks he is shown higher game. When he is five or six months, he is considered old enough to learn how to hunt hares—not at all an easy task, and one which requires a great deal of preparation. The puppy is held in a leash, and led by some of the men to a place where hares are known to lie. The hare is roused and made to run away, and the greyhound is taught to follow it until, after repeated trials, he learns how to hunt it down. When he has thoroughly mastered this lesson, he is promoted to the chase of gazelles, which needs a great deal of caution, especially if the mothers happen to be near by. But in a very little while the pups learn this too, and then the greatest pleasure they have is a hunting expedition.

By this time the pup is a year old, and has nearly reached his full strength and spirits. In three or four months more his heart dances with joy at the sight of a herd of antelopes—the more the merrier, he thinks, as he watches thirty or forty of these big beasts feeding together in the plains. Trembling with excitement he flies to his master, and looks up pleadingly in his face, for he has been too well taught to go off without leave. 'Son of a Jew,' says the master, who himself has discovered the antelopes, and knows quite well what this means—' Son of a Jew, do not lie to me, and tell me you have seen nothing. I know you, friend, and I myself will go with you.' So he takes his skin of water, and sprinkles it over the little greyhound's body, that he may become stronger and better able to resist his enemies. The dog is too impatient to be gone to submit patiently to these ceremonies, and when at last he is set free, gives one rapturous bark, and makes, like an arrow from a bow, for the largest and finest beast in the herd. And when he has killed him, he always receives the flesh off the ribs for his share.

Greyhounds are prudent creatures, but also very vain. If a greyhound fails to bring down an antelope which has been pointed out to him by his master, and another dog succeeds in doing it, he feels wounded in his most tender place. This vanity comes mostly from his education. A pure bred greyhound would never think of eating from a dirty plate or drinking milk which any hand had touched. He learns very early to consider that he has a right to the best of everything. Other dogs may almost starve, and accept thankfully the food a greyhound would not look at, but he will lie by his master's side, and sometimes in his bed. He wears a coat, so that no cold wind may touch him, and if he is cross, everyone declares it is a sign of high birth. No finery is thought too good for him; necklaces and shells are hung round his neck, and he wears a talisman to preserve him from the evil eye. His diet is a matter of careful consideration, and no man would dream of giving his greyhound anything but the dainty bits he has kept for himself.

No well brought up greyhound would ever think of hunting with any man but his master, and indeed his affection and his clean habits amply repay all the trouble spent upon him. If his master is absent for a few days, the greyhound nearly goes out of his mind with joy at his return. He jumps right on to the saddle itself, and almost smothers the man with his caresses. And the Arab understands all he is feeling, and says to him: 'Friend, forgive me, I had to leave you. But now, come with me. I am weary of dates, and need meat, and I know you will be so good as to get me some.' And the dog takes him at his word, for he knows he is worthy of his trust.

When the greyhound dies, the whole tent mourns for him. The women and children weep, as they would for one of themselves, and indeed he is often a greater loss than a member of the family might be. A 'slugui' who hunts for the poor Bedouins is never sold, and only very rarely given away in return for some great benefit. The value of such a 'slugui,' who is a successful hunter of gazelles, exceeds that of a camel; the worth of a greyhound who can capture antelopes is equal to that of the finest horse.

THE LIFE AND DEATH OF PINCHER

PINCHER was a native of Edinburgh, and was born about 1880. It is unfortunate that Dr. John Brown did not write the biography of Pincher, whom he probably knew, while I myself was unacquainted with the hero. This life is based on the recollections of the bereaved survivors of an illustrious hound.

On the mother's side, Pincher came of an old family of fox-terriers. His paternal descent is wrapped in mystery, but those who know the circumstances best believe that Pincher had bull-terrier blood in his veins. His ears were large and loosely flapping; his tail was short, thick, and columnar—that heroic tail which never but once was seen between his legs.

In very early youth Pincher was bestowed on a lady of mature age and maiden dignity, who dwelt in London. She became much attached to Pincher, but soon restored him to Edinburgh. On consulting her friends, and her own sense of propriety, she did not think it becoming that she should constantly appear in police courts. Yet this was her portion in life, owing to the military instincts of Pincher, still uncontrolled by knowledge of the world. Pincher drank delight of battle with his peers, and Wallace rejoiced not more in the blood of Englishmen than Pincher in the gore of English dogs. Through wide Bayswater he kept avenging Flodden, and was in police courts often. He was therefore restored to the bosom of his family, who resided in Douglas Crescent.

Reflection had taught Pincher that a refined Crescent

was no fit arena for military prowess. Besides, he had reduced the dogs of the district to order, and his appearance, like that of the British Flag on the high seas of old, was saluted by tails down. Pincher looked for new worlds to conquer. He took his stand, like some adventurous knight of old, in a pass perilous. He kept that thronged thoroughfare, the Dalry Road, against all comers. No collie, or bull-terrier, or Dandie could pass, but must cross teeth with Pincher. In the Dalry Road he compromised nobody; unrecognised, like the Black Knight at Ashby-de-la-Zouch in 'Ivanhoe,' he reaped his laurels.

Battle was not Pincher's only joy. He loved sacred music. Certain anthems and hymn tunes, when performed on the piano, moved Pincher to an ecstasy which he expressed in rhythmic howls. To secular music he was deaf, or dumb; he did not wed his voice to profane melody. Hence he for long remained apparently indifferent to barrel-organs. But, at last, Pincher was missing from his wonted stand. He kept the pass of the Dalry Road no longer. He had found a wandering musician, proprietor of a barrel-organ, who had the 'Old Hundredth' in his machine. Him Pincher constantly attended in George Square, in Princes Street, in The Pleasance, everywhere. Pincher's family would meet an enthusiastic crowd, who listened with rapt attention while Pincher accompanied the 'Old Hundredth' with vocal and heartfelt psalmody. The musician profited not a little by Pincher's performances.

Pincher could not abide his neighbour, Professor Blackie. The extraordinary liveliness of that scholar found vent in a kind of dance, a sort of waltz in which he indulged as he paced the street. Observing this, and not liking it, Pincher would rush from his lair in the area, circling round the Professor, and leaping up at the tails of his plaid. The learned Professor was obliged to walk like other men in Pincher's neighbourhood.

The Highlands were the home of Pincher's most

celebrated feats, and the Pass of Glencoe witnessed what he doubtless deemed the most tragic event in his crowded life. Here he, who never feared the face of living dog, fled from the dead, as he (erroneously) believed. He was not inaccessible to the terror of superstition, nor could he encounter the foe whom he had already seen stretched lifeless at his feet. But this adventure needs some preface and explanation.

The Coe, after threading the Pass where the massacre took place under tremendous and beetling crags, reaches the sea at Invercoe, above which it is spanned by a bridge. At Invercoe dwelt a family akin to that owned by Pincher. They possessed a Scotch terrier named Jack, between whom and Pincher reigned an inveterate feud. To keep these enemies apart was the great object of all friends of peace. Pincher's family lived on the left, Jack's on the right of the river. One day both families were taking tea in the open air, the table being spread just under the window of a cottage in the village. Pincher was left in the cottage, Jack on the other side of the stream. As the guests partook of the innocent feast, a kind of hairy hurricane sped from above, the urn and teapot were overset, a heavy body landed on the table, and, when the affrighted tea-party recovered the use of their senses, Pincher and Jack were found engaged in a death struggle. Jack, unobserved, had come up the road, Pincher, beholding or scenting him from an upper window, had leaped to the fray!

What could be done was done. Both hounds were lifted from the earth by their tails. Pepper was applied to their nostrils, water was poured over them. But Pincher did not leave his hold till Jack lay motionless at his feet. Then Pincher let himself be dragged off, while medical attendance was called in for Jack, the doctor's house being hard by. The skill and perseverance of that excellent physician were at last rewarded. Jack breathed, he stirred, and, unknown to the relentless Pincher, was conveyed by a band of sympathisers to his own home, very unwell.

After this event Jack and Pincher were carefully kept apart, and Pincher firmly believed that his enemy was dead. But, in the following year, Pincher crossed the bridge, and, in the view of several credible witnesses, he encountered Jack. Instantly that short tail of Pincher's drooped, he trembled, turned, and fled. He had slain Jack, that he knew, and yet here was Jack again, re-arisen from his grave. Now, and never before, men saw Pincher fly from a foe. The inference is obvious: he regarded Jack as a visitor from the world of spirits. Brutus was not afraid of the ghost of Cæsar, but in this one respect Pincher fell short of the Roman courage.

Pincher, though alarmed, was unconverted. Though gentle to small dogs, and the attached friend of little children, Pincher reigned the tyrant of the glen. When he marched down the middle of the village street, dogs and cats fled to back gardens and under beds in cottages. At the age of fourteen Pincher died. It was his habit to jump at the noses of trotting horses; enfeebled by years he 'missed his tip,' was kicked by the justly irritated horse, and never recovered from the injury. Pincher was brave to a fault, tender, faithful, and the patron of at least one of the fine arts: sacred music. When he first landed in the Highlands, the barque which bore him glided through clear water over a green field, submerged at high tide. In the mirror-like expanse Pincher beheld his own reflected shape, conceived it to be a hostile hound, and leaped to battle. His perplexed expression when he rose to the surface is said to have been extremely comic. His old age was gloomy, as he no longer dared to keep the crown of the causeway, dreading the reprisals of the young. The time came to this conqueror when, like Rob Roy in his last days, he had enough of fighting. Such, as drawn by a feeble but impartial hand, were the Life and Death of Pincher.

A BOAR HUNT BY MOONLIGHT

It was shortly before Christmas, when the days are at their shortest, when the sun sets before four o'clock and by five darkness has spread over the face of the land. One such evening there sat smoking and chatting in their comfortable sitting-room the inspectors and the book-keeper of a great estate in Poland, which belonged to a nobleman, but was under the management of a German steward.

'Children,' said the Inspector Wultkiewicz, 'in my rounds to-day I went past the pea-stacks of the Jaguicksy farm. You cannot imagine what havoc the wild boars have wrought there; if it is allowed to go on, by the spring the peas will be completely pulled up.'

At that moment the maidservant entered, and interrupted the conversation by announcing that supper was ready, and all the young men betook themselves to the steward's house across the way, to eat their evening meal in company with the steward's family. At table, the conversation again turning on the wild boars and the damage they had done, the book-keeper declared that in order to drive these pests away for ever it sufficed to shoot one.

Now this book-keeper, who, like the steward, was a German, was very clever at his own business, but, like many other people, believed that he could do everything. For instance, he considered himself an ideal of manly beauty, irresistible to ladies and unsurpassed in all knightly arts. In reality he was narrow-shouldered, hollow-chested, had long spindle shanks and a crooked back, great red hands, and huge feet; he stammered in his speech, and his behaviour towards ladies was like that of a young sporting dog who is being fondled.

'But Herr Vomhammel,' objected the steward, 'you could easily shoot one of them, but you would find it rather dangerous to come to such close quarters.'

'Ah, ha,' laughed Fräulein Anna, the steward's pretty eighteen-year-old daughter, whose chief delight it was to tease poor Vomhammel, 'Herr Vomhammel would take to his heels as soon as ever he heard a pig grunt.'

'Upon m-m-my honour, F-f-fräulein,' he stammered in self-defence, 'I would let f-f-fire with my good revolver, straight on a wild b-b-boar if only it stood still.'

Everyone laughed, but the book-keeper did not seem the least aware of it, and looked round triumphantly.

'An idea occurs to me, however,' said the steward; 'by ten o'clock to-night the moon will be up, so let us invite all the peasants, and especially Ivan Meschkoff, the choir leader, who is experienced in boar hunts, to join us in a raid against them to-night. As you all know, the peasants are not allowed to have firearms, but make use, after the old Ruthenian fashion, of their pikes and pitchforks, once their dogs have brought the creatures to bay. If we were to drive in light sledges through the forest near to the pea-stacks, we might surprise the pigs, and by cutting off their retreat, the dogs that we should take in the sledges with us could not fail to seize some of the herd.'

This proposition met with general assent, the ladies even wishing to join in this sledge expedition, which seemed likely to be free from all danger. The peasants were soon ready, and by nine o'clock assembled, to the number of thirty, under the leadership of the gigantic bearded Ivan Meschkoff, each accompanied by one, or perhaps two, middle-sized but powerful cross-bred dogs, held in the leash. From the village they went to the manor-house, whence all set off together in eleven sledges, as many hunters as spectators. The steward took both his wife and daughter Anna with him on one of the small light sledges, as a large iron-bound one would have dragged too heavily through the freshly fallen snow and have been too wide for the narrow forest tracts through which they had to drive. The steward's double-barrelled gun was the only firearm of any consequence taken on the expedition, and was entrusted to the Inspector Wultkiewicz, as he was an excellent shot. Our friend Vomhammel had naturally not forgotten to take his cherished revolver, with what results remains to be seen.

As they entered the forest, Ivan, who sat by Wultkiewicz on the foremost sledge, desired that all conversation should cease, and soon no sound was heard through the line of sledges but the rattle of the shafts or the occasional neigh of a horse.

Here and there, on either side of the path, where the undergrowth had been slightly cleared, were seen at a little distance numerous shining sparks which might have been taken for glow-worms, excepting for the fact that they moved, or rather seemed to glide along the ground in the same direction as the sledges and at the same rate of speed.

'Are there glow-worms in winter?' asked Wultkiewicz softly of his neighbour.

'No, they are wolves' eyes shining through the gloom,' answered Ivan in a whisper, 'they will follow us to the end of the wood, but there is nothing to fear; they just run with us for their amusement and to see if anything falls off the sledges. They are only dangerous during severe and prolonged cold, for the wolf is cowardly, and seldom attacks but in extreme need, and that never occurs here, with all the roe-deer and wild pig there are in these woods. The peasants and the horses are well accustomed to the sight of the wolves by night, and by day they never appear.'

At a distance of about one thousand feet, the peastacks stood out distinctly against the wintry sky, and all eyes were immediately turned in that direction. In his capacity of leader of the hunt, Ivan gave the order to drive slowly towards the stacks in a large half circle with a gap between each sledge, and to let loose the dogs as soon as the boars should begin to run.

When they had come to within three hundred feet of the stacks, they distinctly saw a large herd of black pigs busily engaged in their work of destruction. As soon as the creatures became aware of the approaching enemy they drew closer together. At a signal from Ivan the greater part of the dogs were let loose, and they rushed barking loudly on the common foe. The whole herd gathered close together in a tangled mass, and took flight across the fields in the opposite direction. Six sledges, manned by pike-armed peasants, pursued them quickly, while the remaining five sledges with the rest of the dogs, drove slowly after them so as to be able to cut off the retreat of the pigs into the forest.

When these last sledges had come to within a hundred and twenty feet of the stacks, the occupants saw a huge dark mass moving among the rooted up straw.

'That must be an old boar, what they call a "hermit,"' said Ivan, 'a dangerous creature that fears neither dogs nor men.'

And, as if to prove the truth of his words, the monster then slowly turned his broadside to the sledge, without interrupting his eating and crunching.

'If only I had my good rifle here!' exclaimed Wultkiewicz excitedly, 'but with shot one cannot pierce a hide like that.'

Vomhammel, who sat on the same sledge beside the driver, no sooner heard these words than he sprang up, threw his long legs over the splash-board, jumped out, and revolver in hand, advanced on the boar with huge strides.

'For Heaven's sake! back! what are you doing?' exclaimed Ivan; but Vomhammel did not heed him, and rushed on.

Wultkiewicz, who did not wish to leave his colleague in the lurch, hastily thrust a couple of cartridges into the

VOMHAMMEL IN DANGER.

barrel of his gun, and hurried to the spot. All these events had taken place with the speed of lightning, and in the general surprise every one stood helpless, Ivan alone not losing his presence of mind.

'Wasil!' he cried to the driver of the sledge, 'drive quickly forward! and let loose the dogs.' And immediately eight large dogs sprang to earth.

Meanwhile Vomhammel had approached to within sixty feet of the boar, then he stopped, took aim, and fired three times in rapid succession, without any shot, however, taking effect. Slowly the monster raised his great broad head at the noise, then at sight of the disturber of his peace he gave vent to a series of grunts, and struck his mighty tusks on the ground. Vomhammel's courage instantly vanished, and, letting fall his revolver, he quickly ran back. 'Here! here!' called Ivan, and with such strides as never were seen, Vomhammel made for the rapidly approaching sledge. But the boar was as quick as he, and apparently meant to avenge himself for the insult done him. Lowering his head, he rushed after the flying enemy, ploughing up the snow with his tusks. Soon he was close upon him, and Vomhammel seemed lost, as there was still a considerable distance to cover before he could reach the sledge. Just then a shot rang out, and the boar fell forward. Wultkiewicz had fired a shot at him from a distance of about thirty feet. Immediately the boar was on its feet again, though limping on a fore leg; the short delay, however, had been enough to save poor Vomhammel. As the boar, blinded with rage, hurled itself against the sledge its victim's long body was already safe, only his legs hanging down on the wrong side. A blow from the boar's tusks hitting one of those long lank limbs, ripped up the boot from top to bottom. Ivan, with his powerful left hand, firmly grasped Vomhammel's body, and thus rescued him from further attack, while with the right he dealt the boar a spear thrust. The dogs also flung themselves on the monster, which was attacking the sledge so furiously that it certainly would have been overturned but for good driving. When at last the remainder of the party appeared on the field of battle, the boar, after a hot struggle, had been completely vanquished, some pikes had been bent and broken in the combat, and one dog had paid for his valour with his life. The other hunters had a large sow to show as their spoil, which they had succeeded in slaying without any mishap to themselves.

When all had reassembled on the scene Fräulein Anna's eye fell upon Vomhammel, who lay all huddled up on the sledge. When she saw the gaping rent in the boot, she exclaimed : ' See, Herr Vomhammel is terribly wounded ! '

Everybody ran to look, but found after all that the boar's tusks had torn nothing but cloth and shoe-leather.

' God be thanked,' exclaimed Anna, ' that it is no worse ! '

Vomhammel had the satisfaction of being for some time after that the hero of the day, and his beloved Anna has never again twitted him with lack of courage.

THIEVING DOGS AND HORSES

It is now about eighty years since Sir Walter Scott told some curious stories, proving how animals could be deliberately trained by their owners to break the law, or to help them to break it, all the while thinking they were acting from the best motives, and only doing their duty. It is, if we come to reflect, very difficult for a dog to learn that he is worthy of praise if he defends his master's property, while he is doing a very wicked thing if, at that very master's bidding, he tries to get possession of somebody else's. His only idea of the whole duty of dog is to do what he is told. And a very good idea it is, too, only it sometimes leads to trouble. Why, only a few days ago a large boar-hound was trained by some Paris thieves to fly at a man's throat at a given signal. The man was nearly killed, but not before the dog and his owners had been caught by the police. The thieves were taken to prison, and the dog to the lethal chamber.

This little incident shows that the nature of dogs, as well as that of men, is pretty much the same as when Sir Walter was writing about them. Somewhere about the year 1817 a constable made a complaint to the police magistrate of Shadwell, a large district in the East of London, that a horse in the neighbourhood had become a confirmed hay-stealer. Every night, declared the constable, that horse would walk boldly up to the stands of hackney coaches in the parish of St. George's-in-the-East, and eat as much hay as he wanted, after which he instantly galloped away. More than once a party of men had set out to catch him, but in the end they had been obliged to give this up, for if they attempted to interfere with him when he was eating, he would first turn round and charge them, and then kick furiously at them; and if this did not do, he would end by biting them. So, not knowing what to do, they had sent the constable to the magistrate to ask his advice.

It was not of much use when he got it. The magistrate thought it was a very shocking state of things, and directed that the offending horse should be brought into court to answer these grave charges, *if he could be caught;* but this was exactly the difficulty, and as there is no record at the Shadwell Police Court of the case being tried, it is probable that one of two things happened: either the horse was shot by one of the angry drivers, or he went on stealing hay as long as it pleased him.

The next time we hear of a four-footed robber being charged in a police court it is at Hatton Garden, a part of London that is inhabited by Italians and diamond merchants, and on this occasion it was a dog who was the thief. Two ladies appeared one morning before the magistrate, and one of them stated that as she and her sister were returning from St. Pancras Church the evening before, and were walking down the road to Battle Bridge about six o'clock, a hairy dog, not unlike a collie, had suddenly jumped up from the roadside where he had been lying in wait, and seizing a small bag (or reticule, as it was called in those days) which one of the ladies held in her hand, dashed off with it across the road, and was lost to sight in the darkness. Her loss was heavy, for she was not rich, and the reticule contained a sovereign, eighteen shillings in silver, a silver thimble, a pair of silver spectacles, and two or three other small things. Perhaps she had been spending the afternoon at one of the little card-parties which at that date had hardly ceased to be the fashion.

When she had told her tale, a constable came forward and stated that, only the Saturday before, a dog answering to the same description had attacked a poor woman in the neighbourhood, and snatched from her a bundle containing two shirts, some handkerchiefs, and other articles of dress, and had run off with them, leaving the woman so frightened that she had nearly died of terror. And these charges were not the only ones that were lodged against this dog. Four or five more complaints of robbery were brought against him, and though no man had ever been seen in his neighbourhood, at the time the thefts were committed, it was supposed that he must have been carefully trained to the work, and also to bring his spoil back to his master, who would be hiding in some place not far distant. In the end, the constable undertook to stop his pranks, or else to shoot him.

Sometimes, however, it is not possible for the master (and real offender) to keep entirely in the background, and instances have been known of the punishment falling on the right head.

Towards the close of the last century two men and a dog were tried for sheep-stealing before one of the most celebrated Scotch judges of the day.

One of the men, Murdieston by name, lived on a farm on the north bank of the Tweed, nearly opposite the beautiful old castle of Traquair; the other, who was called Millar, was his shepherd. They were much respected by their neighbours as quiet industrious people, but in reality had carried on the business of sheep-stealers for many years without exciting the suspicion of any one. Indeed, they were so very cautious that, even in the middle hf the night, they would never drive the stolen animals along the high road, lonely though the country was, but preferred to keep to the side of the bare hills that lie between the little river of Leithen and the Tweed. Not that they were safe even here, for a careful shepherd would often make the round of his flocks by night, or it would happen that the sheep gave more trouble than Millar expected, and precious time was lost, so dawn would come while the farm was still many miles away. Then he would make his way to the bank of the river, which lay in an opposite direction, and leave his dog Yarrow to bring the sheep back to the ground belonging to Murdieston, where they would be quite safe from suspicion if any one passed by.

A short distance from the river was an old square tower, to which the farm-house had been afterwards added, and under the tower was a large cellar, where the stolen sheep were generally concealed. On Sunday mornings, when everybody was off to church, the thieves busied themselves with changing the marks that are always put upon sheep, and replacing with their own those of the real owner. During this operation Yarrow kept watch outside, and never failed to give a warning bark when he caught sight of a stranger on the road or on the hill.

Of course Millar knew quite well that if he went on keeping his robberies to one district he would certainly end by being found out, and that before very long. So, one night, he crossed the Tweed to a lonely farm in the hills of Selkirk, where he managed to get hold of several sheep, and prepared to drive them home. Now sheep have a strange objection to coming down a hillside at night, and still more to crossing a river; so, when Millar, after steering his flock with some difficulty round the shoulder of Wallace's hill, tried to induce them to swim a pool of the Tweed, the elder members of the party became obstinate, and stubbornly refused to budge one inch. It was to no purpose that Millar and Yarrow did everything they could think of to force or persuade. Across that river they would not go, and, to his despair, Millar saw the day breaking over the east, and knew that he must fly at once, if he did not wish his own neck to be in danger. Yet he could not bear to give up his booty just at the last, when he was hardly a quarter of a mile from the tower, so, leaving Yarrow in charge, he went home, calling directions to the dog as long as he dared.

Left to himself, and feeling that he was put upon his honour, Yarrow rushed furiously at the oldest and most obstinate ewe on the ground, and drove her into the water, frightened out of her wits, for she thought she was going to be bitten; struggling to get away, two others tumbled over the bank after her, and were drowned in the stream; the rest became wilder than ever, and as by this time the sun was well above the horizon, Yarrow knew that he too must follow his master, and leave the sheep to their fate.

Late that same evening the sheep might have been seen wending their way wearily home with new marks on their bodies, hastily daubed on by Millar in a lonely hollow of the hills.

The thieves thought that they had escaped before any prying people were up and about; but they must have been watched by some unseen person, for information of their misdoings was given, and they were soon lodged safe in gaol. The case was easily proved, and both Millar and his master condemned to death, for in those days there were very few crimes which did not lead to the gallows. When he saw that it was useless to deny the fact any more, Millar told the whole story to a respectable sheep farmer who came to visit him in prison, and they both agreed that they did not know which was most surprising, the obstinacy of the sheep in refusing to cross the river, or the perseverance of the dog in trying to force them to do it!

The two thieves were hanged on the appointed day; but Yarrow was bought by a sheep farmer in the county, who hoped to train him to honest work. But it was too late; his teaching had all been in one direction, and when he found he was not allowed to show his cunning in driving away other people's property, he grew quite stupid, and could never be trusted to do even the commonest everyday tasks which fall to the lot of every collie.

However, it is not only collies which can be taught to steal, though, of course, dogs are like children, and some of them learn much more quickly than others. Some years after Millar's bad conduct had met with the reward it deserved, a rich young man, living in Edinburgh, saw a beautiful and clever little spaniel which took his fancy, and he never rested until the owner had agreed to sell it. The animal had been in his new home only a few days when its master was astonished and shocked at its bringing home a pair of new gloves, three silk handkerchiefs, and, shortly after, a lady's gauze scarf. At first he tried to believe this was an accident; but as the collection grew larger and larger, he soon understood that thieving had formed the largest part of the dog's education, and that most likely it would be quite impossible to cure the animal of its bad ways, now that it had grown up.

So, when the spaniel next began whining and sniffing at the door, and showing all the usual signs of wanting to go out for a walk, the young man took down his hat, and turned into the streets, watching all the while what his dog was doing, though very careful never to turn his head in that direction.

And what the dog did was very curious to see. It loitered through the town in the purposeless way that all dogs think is a proof of gentlemanly behaviour, stopping every now and then either to speak to a friend, or to examine something strange that lay in the gutter. The young man walked steadily on, and entered a shop where he was well known, telling the shopkeeper, hastily, to take no notice if the dog should enter, as he would of course pay for any of its robberies. He then began to turn over some of the articles for sale, so that the animal's suspicions might not be awakened if it came in, which it presently did, in the same lounging, careless manner that had marked its walk through the streets, treating its master as if he was a person whom it really was not respectable to know. While the spaniel was thus poking round the shop with its eyes apparently turned in another direction, the young man was turning over some articles at the counter. Suddenly he glanced at the dog, touching, as if without thinking, a small parcel that lay there. Soon after he left the shop.

The dog, who from first to last had given no sign that it and its master knew each other, sat down peacefully at the door, in a position where it could see all that was going on inside. At length the shopkeeper went for a moment to an inner room to fetch something he wanted. In an instant the spaniel had placed its fore-paws on the counter, seized the parcel, and crept out noiselessly to rejoin its master, bearing the stolen property triumphantly in its mouth.

We are not told whether in the long run the young man was ever caused any serious trouble by this magpie of a dog; but a gentleman who became famous as a lawyer at the end of the eighteenth century very nearly fell a victim to the too faithful memory of his horse.

In the days of his youth, somewhere between 1750 and 1760, the journey between Edinburgh and London was made on horseback. If a man was rich enough he hired horses to meet him and his servant at certain places on the road, but if he was poor, he bought a horse at the beginning of his journey, and sold him for what he could get at the end of it.

Now this gentleman had been brought up in the country, and nobody was a better judge of a horse, so when the business which had brought him to London was finished, he set out for Smithfield, where the great horse market then was, to buy a mount for his return journey next day. He instantly picked out a handsome creature with a beautiful head, and stopped to look at it, though he felt, with a sigh, that the sum asked would be certain to

THE HIGHWAYMAN'S HORSE

be far higher than he could afford. The horse dealer, however, at once came up, and, while praising the horse, named such a low price for it that the gentleman could hardly believe his ears, and made sure the animal must have some serious drawback. He examined it carefully all over, but could find no drawback anywhere—it was beautifully proportioned, and its knees were quite sound. The dealer, mistaking the reason of his silence, was so anxious to have the bargain concluded that he agreed to accept a still smaller sum, and the young man, feeling that there was some mystery somewhere, paid the money down, and the following morning took the Great North Road to Edinburgh.

For the first few miles out of London the way was full of people, and no man was better mounted than himself, or had a horse with better paces. In fact, the more pleased the young man got, the more puzzled he became. As they approached Finchley Common the number of riders fell off, and by the time the young man reached a dip in the road not a soul was in sight but a clergyman driving a one-horse chaise, which was travelling in the opposite direction. As they came close to each other, the ridden horse stopped dead in front of the driven one, thus preventing it from going on its way. The clergyman, taking for granted that he had to do with one of the highwaymen who in those days were the terror of every country district, quietly got out his purse, and assured the young man, who all this while was speechless from astonishment, that it would not be necessary for him to use force. The shame caused by this remark loosened the rider's tongue, and, with a hasty apology and a confused explanation, he whipped up his horse and went his way.

On the next occasion, however, that the horse thought fit to exercise the profession for which he had been educated, things took a graver turn. This time he halted in front of a coach, and before his rider knew what he was at, or was able to get him under way again, a blunderbuss was aimed at the poor man's innocent head, and he was informed that the occupants of the coach would sell their lives dearly! And, as if this was not enough, it appeared that the horse was well known all along that very road, and when he had escaped from the firearms of passengers, it was only to be stopped by the officers of the peace, hoping at last to capture the notorious highwayman who had so many times contrived to slip through their fingers.

It can easily be imagined that by the time York was reached the poor young man had had quite enough. He parted with his prize for a mere trifle, less even than what he had paid, and was glad to buy for a much larger sum a horse that was not indeed so handsome to look at, but had been better brought up.

Yet it would be unjust to think that an animal's misdoing is always the fault of its master and mistress. Here and there we find a creature who is naughty or tiresome just because it likes it, and who will not suffer itself to be taught better ways. Not long ago a dog was living at the mouth of a short street in London which was open only at one end, and was the home of a great many children. If any child tried to pass him he would run at it and snap, and if he did not actually bite them, the nurses always thought that he had done so. At length every one became so frightened that the father of one of the little girls had to go before a magistrate and beg that the owners of this terrible animal might be forced to get rid of him, as it was not fair that the whole street should be kept in a state of siege, only for the amusement of one dog. The magistrate agreed that it certainly was unreasonable, and from that day the children could come and go as they chose, without any fear of suddenly being sent sprawling on the pavement.

TO THE MEMORY OF SQUOUNCER

Squouncer was a dog by himself. Other dogs may boast of belonging to large families of collies, greyhounds, or dandies, with cousins as numerous as the sands of the sea; but there could only have been one Squouncer.

How did he get his name? Well, his master (before he *became* his master) saw the word Squouncer in a book he was reading, and thought it so delightful that he instantly made up his mind to search through the world till he could find a dog that would fit it.

And one day he found Squouncer. What was he like? He was what the French call a 'Beau-laid'—'beautiful-ugly.' His ancestors may have been bull-dogs, and it is whispered that they gained their laurels in Spain. Squouncer was a middling-sized dog, with a golden-brown skin, much the colour of dark amber. And he had a broad face, and a nose which stuck out that gave him the air of what used to be known as a 'fire-eater.' Like another gentleman of a similar disposition, he might have been nicknamed 'fighting Bob' if you had only gone by his looks, but a milder-mannered dog never snorted when he breathed—as long as there was no food in sight. Then, all the lion in Squouncer's forefathers rose up, and woe be to the person who came in his way.

It was just because he was so different from any other dog that ever was or ever will be that his master and mistress were so fond of him. Anybody who reads history or has his eyes open can see that it is not the good people or the handsome people that have really been loved most and remembered longest, but the people who have made us laugh! Why, even the most wicked and gloomiest kings had their jesters, and often the jesters were able to tell the kings very disagreeable truths, or to beg off some poor wretch condemned to death, when a word from any one else would simply have sent him to share the fate of the criminal.

Now it may be doubted whether, even if he had had two legs, and had lived in the palmy days of long ago, Squouncer would ever have interfered to snatch people from the gallows. He was not (except where his food was concerned) a very courageous dog, and he never *could* make up his mind what he wanted to do, what he ought to do—and no one that goes through life on these principles will ever be a hero. Sometimes his master and mistress used to amuse themselves with this weakness of his. They would sit at each end of a long room, and one would call 'Squouncer.' Squouncer, who had very early been taught to come when he was called, rose at once and started to obey. 'Squouncer,' said a voice behind him before he had got half way. He stopped, listened, and turned slowly round. 'Squouncer' was again repeated from the further corner; and poor Squouncer halted again, and looked piteously from one to the other, but never thought of doing the only sensible thing, which was to lie down before the fire and pay no attention to anybody.

One dreadful day, a young black retriever suddenly appeared in the house. There ought to have been nothing disturbing in this, as the animal was friendly and playful, and quite ready to be polite to Squouncer—who was an older dog than he. But Squouncer's thoughts at once flew to dinner-time, and so did his master's and mistress's, and they determined to watch and see what would happen.

And what did happen was this. The two large tin plates were placed side by side in the tiled hall, each filled with a delicious mess enough to warm the heart of any dog. And not only his heart: for if you had once looked at Squouncer going to his dinner, you would have had no difficulty in understanding the expression 'your eyes starting

out of your head.' Well, Squouncer dashed straight at his plate—the biggest you may be sure, and the fullest—and gobbled up the contents so fast, and with such a disgusting noise, that the tin plate performed a kind of dance all round the hall, Squouncer's tongue never leaving it as long as the tiniest scrap remained to eat. When it was as bare as Mother Hubbard's cupboard, he left it lying where it was, and pushing the retriever (who was taking his dinner in a polite and gentlemanly manner) rudely to one side, he began the same game over again. The retriever was so astonished at this behaviour that he meekly stood back, and before he had collected his senses, the second plate was as bare as the first. Then Squouncer's master thought it was time to interfere, and took the retriever off to the kitchen, where he might eat his food in peace.

This success was very bad for Squouncer, for it made him despise his new companion, and think he could treat him as he chose. For several days he continued to swallow his own dinner with the same noise and indecent haste, so as to secure the best part of Negro's. He did not even take the trouble to be pleasant to him between whiles, and when one afternoon, after a huge meal, Negro detected him secretly burying some pheasant bones under a tree till he should have recovered sufficient appetite to eat them, the retriever's temper gave way, and he resolved he would stand this sort of thing no longer.

So the following day at two o'clock, when the plates were put out for dinner, and Squouncer's tin plate was heard as usual rattling round the hall, pushed over the tiles by that long, greedy tongue, Negro cocked his ears and made ready for battle. Suddenly the noise ceased, and a second later he was almost thrown down by a violent push as Squouncer advanced to the charge. What occurred next was never clearly known to any one; but a frightful shriek brought every one into the hall, where a black and yellow ball was rolling about wildly. The black half was uppermost, and was hauled off by his master, and then Squouncer's leg was found to be broken. Poor Squouncer! he never recovered the shock and the shame of that fight. He was so unhappy at the sight of his conqueror that his mistress took pity on him, and gave Negro to some friends. After a while the broken leg mended (though it left a limp behind), and Squouncer's appetite was found as healthy as ever. He lived many years; and his death, in a good old age, left a blank in the house. A black Spanish bull-dog now reigns in his stead, which may have its virtues, but will never be half as good company on a wet day as Squouncer.

HOW TOM THE BEAR WAS BORN A FRENCHMAN

CAPTAIN PAMPHILE had made many voyages in southern seas, and traded in gold dust, spices, and ivory; so he thought that the north might be a pleasant change, and that he could do a little business in furs and train-oil.

Now this happened more than sixty years ago, and the voyage took longer than it would in our days. And when at last they reached land, the Captain thought he would take a holiday, and go on shore for sport, leaving the ship in charge of the chief mate.

He plunged inland at once, and after some days' march reached a great forest, where he hoped to find game; but as night came on, he realised that he did not know his way. It was not a cheerful prospect, for his clothing was light, and many growls were heard around, amongst which he recognized the voices of the hungry wolves abounding in these forests. He looked round for shelter, and chose a sturdy oak, which he climbed—only just in time, for the wolves, who had scented him from afar, came hurrying up in hopes of a good supper. But they were too late; the Captain had found a perch!

But the wolves hoped on, and huddling round the tree, moaned and howled so fearfully that the Captain could hardly restrain a shudder. Through the darkness he could still trace the outline of their shaggy backs and catch the gleam of their fierce eyes. This constant watch made him almost giddy, and, fearing a fall, he tied himself firmly to a bough with a rope he had with him. Then he gripped the branch overhead and closed his eyes.

Soon he became drowsy, and had a strange dream. A whistle seemed to sound overhead and something chilly to be stifling him with great coils. This gradually passed, and the ghosts of wolves seemed to fade and their howls to decrease as the tree bent and rocked; then all was silence.

After this the Captain fell sound asleep, and did not wake till dawn. As he opened his eyes the first thing he saw was the green boughs overhead through which were glimpses of blue sky. Then he looked down, and at once the terrors of the night were explained. The ground all round the tree was scratched up by the claws of many wolves, whilst one of them, crushed almost out of shape, lay there half swallowed by a huge serpent whose tail was still coiled round the tree.

The Captain trembled when he saw the double danger he had been in: the wolves at his feet and the serpent overhead; for he remembered the whistling sound, and the clammy folds which had so nearly choked him. He remained for some time staring at the strange sight before him, but at last dropped carefully to ground, and hurried away as fast as his feet would carry him.

There was no road in the forest, but the hunter's instinct, combined with the sailor's science, soon enabled him to strike on a track through the thick vegetation. He was hungry, but as, in his haste to fly from the wolves, he had lost his gun, all game was beyond his reach, and he had to be content with such roots and berries as he could find.

At length he thought he saw daylight more clearly, and, quickening his steps, arrived shortly at the outskirts of the forest in sight of a great green plain with a line of mountains beyond. To his joy a thin column of smoke in the distance gave signs of some habitation, and he struck at once towards it.

It had grown dark before the Captain reached the hut from which came the smoke. As he drew near he saw that the door stood open and that a bright fire burnt on the hearth within. Before the fire the dark shadow of a woman passed to and fro.

He paused on the threshold, asking leave to enter, and on receiving an answering grunt, he stepped in and drew up an old stool near the fire. Opposite him crouched a young Sioux Indian, holding his head between his hands and seeming deaf to any sound of the stranger's approach.

Pamphile looked at him, wondering were he friend or foe.

'Does my brother sleep?' he asked at last.

The Indian raised his head and pointed to one of his eyes which had evidently just been shot out by an arrow. The Captain asked no more questions, but turning to the old woman said: 'The traveller is tired and hungry; can his mother give him food and shelter?'

'There is a cake under those ashes and a bearskin in yonder corner,' replied she. 'My son can eat the one and sleep in the other.'

'Have you nothing else to eat?' inquired Pamphile.

'Oh, yes; I've got other things,' said the crone, fixing a longing gaze on the Captain's watch-chain. 'I have— that's a fine chain of my son's—I have salted buffalo and some good venison. I wish I had such a chain.'

'Well, well, bring me some meat,' replied Pamphile, avoiding a refusal; 'and have you a bottle of corn brandy by you?'

The old woman raised a partition of matting and disappeared into the inner part of the hut. As she vanished the Indian raised his head.

'Does my brother know where he is?' he asked the Captain.

''Pon my word, no,' was the careless answer.

'Has my brother any weapon with which he could defend himself?'

THE CAPTAIN HAD A STRANGE DREAM

'None.'

'Then let him take this knife, and be careful not to go to sleep.'

'And you?' asked Pamphile, hesitating to accept.

'I have my tomahawk—silence!'

So saying, he dropped his head between his hands again and became immovable.

The old woman raised the matting and brought in supper, and the Captain slipped the knife into his belt.

The woman's eyes turned to the chain once more.

'No doubt,' said she, 'my son met some white man on the war-path. He slew the man and took his chain.'

'You are mistaken, mother,' said the Captain; 'I have been hunting buffalo and beaver as far up as Lake Superior; then I took the skins to the town and changed half for this watch-chain.'

'I have two sons,' remarked the woman, placing the supper on the table. 'They have hunted these ten years, but have never managed to get such a chain as that. My son said he was hungry and thirsty; let him eat and drink.'

'Does not my brother of the prairie sup?' asked Pamphile, drawing his stool to the table.

'Pain stops appetite,' was the reply. 'I am not hungry, but I am weary, and going to sleep. May the Great Spirit keep my brother.'

'How many skins did my son give for the chain?' began the covetous woman.

'Fifty,' said Pamphile at haphazard, falling to on his supper.

'I have ten bear and twenty beaver skins here. I will give them for the chain.'

'The chain is fastened to the watch,' replied the Captain. 'They cannot be separated, nor do I wish to get rid of them.'

'It is well,' said the woman with an evil smile. 'Let

my son keep them. Every living man is master of his property; only the dead possess nothing.'

The Captain glanced hastily towards the Indian, who did not move, and fell to on his supper as heartily as if he feared no danger. When he had finished he threw himself on the buffalo skin, but with no idea of going to sleep.

He had not been lying down very long, when the matting was raised and the woman peeped in cautiously. Neither sleeper stirred, so she went to the door of the hut and listened. No one was in sight, and she turned back and began to sharpen a long knife. The Captain watched her through his eyelashes and drew his own knife from his belt, opened it, and felt the edge.

Then steps were heard, and a minute later two big young men appeared bearing some game. They paused to look at the sleepers, and one of them asked his mother how they came there. For reply she led them silently behind the partition.

The Captain noiselessly turned so as to face the young Sioux, and noticed that, though apparently sound asleep, his head rested only on one hand, while the other lay by his side near his tomahawk.

Just then the matting was raised and the young men crawled silently under it, their mother's head just peeping out behind them.

Each approached one of the sleepers, then paused, looking at their mother.

'They sleep,' she whispered; 'go on!'

At her word each son raised his arm to strike, but instantly fell back with a cry.

The Captain had plunged his knife into the breast of one, and the Sioux had split the skull of the other.

The old woman uttered a despairing shriek, and rushed off to the forest, and the Indian, picking up a lighted brand from the hearth, proceeded to set fire to the hut, whilst he executed a triumphant war dance round it.

Then he turned to Captain Pamphile: 'Where does my brother wish to go?' he said.

'To Philadelphia.'

'Follow me, then,' and the Indian strode towards the forest.

They walked all night, and at daybreak came to the open plains. Here the Indian halted.

'My brother has arrived,' he said; 'from the top of that mountain he will see Philadelphia.'

With these words he plunged back into the forest, and the Captain set out to climb the mountain.

On reaching the top he found his guide had said true, and he saw Philadelphia lying between the green waters of the Delaware and the blue waves of the ocean. Off he started in high glee, though his goal looked quite a two days' march off. He was stepping briskly along, humming a tune and swinging a stick, when he noticed a black object at some distance. As he drew nearer the object seemed to approach too, and at length he made it out to be a negro.

This was lucky, for he wished to find some place where he could sleep. So he hurried on till he was face to face with the person he had seen.

Then he discovered his mistake. It was not a negro but a bear!

The Captain saw his danger, but he did not lose his presence of mind—though a glance round showed him there was no means of escape.

The bear on his side halted some yards off and examined the Captain.

The Captain reflected that many bullies are cowards at heart, and that possibly the bear might be as much afraid of him as he was of the bear; so he advanced.

The bear, not a bit frightened, advanced too.

The Captain turned on his heels to retreat, but after three steps found his way barred by a rock. Leaning against it, so as not to be surprised in the rear, he waited.

He had not long to wait. The bear, a huge animal, followed exactly in his footsteps and marched straight upon him.

The situation was unpleasant, for the Captain's only weapon was his stick, and when the bear arrived within two paces of him, the Captain raised it. The bear instantly rose on his hind legs and began to dance!

It was a tame bear, which had broken loose and escaped.

Captain Pamphile, reassured by his enemy's deportment, now noticed that he was muzzled, and that part of the broken chain still hung at his neck.

He at once saw all the advantages to be derived from such companionship, so, seizing the end of the chain, he resumed his journey, leading the bear like a dog.

Towards evening, as they were crossing a great field, he noticed that the bear tried to stop near certain plants, which were unknown to him. Thinking there must be some special reason for this, he made a halt the next time it happened. The bear began to claw the ground and grubbed up a number of tubers or roots. Pamphile tasted one, and found it excellent, with a flavour reminding him of truffles. This was a valuable discovery, so he let the bear continue his hunt, and in an hour they had collected an ample supper for man and beast.

Then the Captain took note of a tree standing by itself, and having carefully examined it without discovering the trace of any reptile, he tied his bear to the trunk, used his back as a stepping-stone to the branches, and soon made himself a bed, where he slept soundly all night.

Next morning he woke refreshed and saw the bear sleeping quietly below. He climbed down and roused him, and both marched on so briskly that they reached Philadelphia by eleven o'clock that night.

Here a fresh difficulty arose. No innkeeper cared to house a savage bear at such a late hour. One after

THE BEAR INSTANTLY ROSE ON ITS HIND LEGS AND BEGAN TO DANCE

another refused, and our Captain was beginning to despair when he spied a brightly lighted inn from which came sounds of singing and laughing. He felt certain some ship's crew was making merry within, and pressed on, when the sound of a well-known song of his own country caused him to stop suddenly. He listened, with a heart beating for joy, and waited for the chorus. Yes, he was right! These were his fellow-countrymen, and, looking through the open door, he saw that they were not only his countrymen, but the crew of his own ship the 'Roxalana.'

He did not hesitate an instant. Thanks to his privations and adventures he was hardly recognisable. He pushed the door wide open and walked in, followed by his bear.

A loud cheer greeted them.

The Captain regretted not having held any rehearsal, but the bear took the whole matter into his own hands —or paws.

He began by trotting round to clear a circle. The sailors stood on the benches, the first mate seated himself on the top of the stove, and the show began.

Everything a bear could learn that bear knew. He danced the minuet, rode astride on a broomstick, and pointed out the greatest rogue in the company. Such a shout of delight greeted the end of the performance that the mate offered to buy the bear as a present to the crew.

The Captain accepted the offer, but slipped out at the beginning of the second part of the performance without being recognized by any one.

He made his way to the harbour, and after some time succeeded in finding the 'Roxalana,' swung himself on deck, and went quietly down to his cabin and to bed.

The crew came on board much later, and what was their surprise next morning to see their missing captain appear among them and take command of the vessel as if he had never been absent.

He at once set sail for France, taking the bear with him, and as soon as possible after reaching port he set out for Paris, intending to present his capture to Monsieur Cuvier, the naturalist.

Just after reaching Paris, the bear gave birth to two little cubs, and the Captain, always pleased with a good bargain, sold one to the hotel-keeper, who sold it again to an English gentleman by whom it was brought to London.

The other cub was sold to Alexandre Décamps, who named it 'Tom,' and confided its education to his friend Fau, with the admirable results we have read of in the 'Blue Animal Story Book.'

(Adapted from A. Dumas.)

CHARLEY

It is nearly seventy years since an American, named Catlin, set out from his home in Wyoming to travel westwards through the great country which was then only inhabited by Indians and wild animals, but is now full of flourishing towns.

In the course of one of these journeys Catlin fell very ill, and it was many weeks before he was fit to leave his bed. When, however, he got better again, he sent for his horse, Charley, which had grown fat on prairie grass —and looked very unlike his master—and the two prepared to start for the River Missouri, more than five hundred miles away.

Catlin's heavy luggage was sent by steamer to meet him at St. Louis, on the Mississippi, but there was still a good deal left for Charley to carry. A bear-skin and a buffalo robe were spread across his saddle, a coffee-pot and a teacup were tied to it, a small portmanteau was fastened somewhere else, and in the portmanteau was a supply of hard biscuits; while Catlin sat in any space that was left, with a little compass in his pocket to show him which way to go, and a gun and a pair of pistols in his belt, in case man or beast should attack him.

So day after day Charley and his master rode on toward the north, through plains of grass all covered with flowers. Every night when the sun set Catlin jumped off and unloaded his horse, which he tied up, or 'picketed,' with a long rope, so that Charley should have plenty of room for feeding. Then he lit a fire to keep off the wolves, whose snarls and howls were always to be heard in the distance, and lying down on his bearskin with his saddle for a pillow and his buffalo robe for a blanket, Catlin curled himself up, and slept soundly till morning.

One evening Charley was picketed as usual, and his master had gone down to the banks of the stream to get some water for his coffee, when the horse, being in a mischievous frame of mind, slipped his rope, and went off towards a patch of grass, which he thought looked much greener and juicier than what he had been eating. Catlin soon saw what had happened, and picking up the lasso with which wild horses are always caught, he started after the runaway.

But it was no use; Charley knew all about lassoes, and exactly how far it was safe to let them get near you. Besides, he wanted to have a little fun and to tease his master, and each time the lasso was thrown Charley was always just a tiny bit out of reach. It soon grew too dark even to see where the horse was, and as Catlin was still weak from his illness, and easily tired, he gave up the chase, and stretching himself out before the fire, made up his mind that he would have to finish his journey on foot.

It was the middle of the night when he woke with a start, feeling some huge creature bending over him. An Indian, of course, which had tracked him while he slept, and had followed him to take his scalp! For an instant the poor man's heart stood still; then a soft nose touched him—and the Indians' noses are not soft, and they are not much in the habit of rubbing people's faces with them! It must be Charley after all, and Charley it was, standing with his fore-paws on his master's bed, his head nodding in a sound slumber!

As soon as Catlin got over his fright he went off to sleep again, and did not wake until the sun was well above the horizon. His first thought was for Charley,

THEN A SOFT NOSE TOUCHED HIM

and great was his relief to see that naughty animal having breakfast at the edge of the stream, looking as if no idea of running away had ever occurred to him! But he was still inclined for a game, for when his master, after a hasty meal of coffee and biscuit, came down to the stream to catch him, Charley danced about a little more, taking care just to keep out of reach.

At last Catlin thought he would try what a trick would do, and flinging the skins round his own body, and the saddle over his back, he began to walk away. For a quarter of a mile he tramped on steadily without once looking round, then he took a hasty glance over his shoulder. Charley was standing quite still near the fire, which was still burning, watching his master. Suddenly he went straight up to the place where Catlin had slept, and finding nothing there, threw up his head and neighed loudly. In another moment something rushed wildly past Catlin, who was walking steadily on, and, wheeling sharply round, stood trembling before him.

Catlin took care not to do anything which might startle the penitent, and called him gently by his name. But Charley had had a fright, too, and had no longer any wish to play with his master. So when Catlin drew near him with the bridle in his hand, he actually bent his head to receive it, and remained perfectly quiet while the saddle was being fastened on his back.

All through that day they journeyed on over the prairie, with its endless waves of flowery grass, and late in the afternoon arrived at a beautiful little valley, where Catlin determined to pass the night—and *this* time he determined he would run no risks with Charley. A clear stream ran through a smooth lawn, and *in* the stream were fish, and *on* it was a brood of fine young ducks. Large trees were dotted over the smooth grass, and between the wild plum and cherry trees laden with fruit hung vines bearing clusters of purple grapes. Underneath, the ground was bright with sunflowers and sweet with lilies and violets.

No place could be more lovely and peaceful; and after making a hearty supper of perch and broiled duck, Catlin went for a stroll to explore a little further.

Five hundred miles is a long way to ride, and in the course of his journey through the prairies Catlin had to cross several of the big rivers which run into the Missouri or the Mississippi. This was not always easy, for there were no bridges to be found, and the streams were often both deep and rapid. There was also another danger to be feared, besides that of being drowned or falling a prey to the Indians; and this was the very deep ditches or sunken streams, with their tops entirely hidden by the long grass, into which a horse might suddenly fall and injure himself and his rider. After a while, Catlin learned to be on the look-out for these pitfalls, and to know their signs, and as they had to be crossed somehow, there was nothing for it but to go at them boldly, and to trust to luck to getting out again. This was generally a very difficult matter; the streams were often full of mud, and till you were in the middle of them you had no notion how deep they were, and not always then. On one occasion Catlin had ridden along the edge of a stream of this kind, in order to find a ford, but as this seemed hopeless, he plunged in at a place where it was six or eight yards wide—as to the bottom, they never touched that at all. They made for the bank, which was of clay, and rose straight out of the water. Catlin managed to catch hold of the top and drag himself up, clutching Charley's bridle in his hand; but he saw directly it was quite impossible for the poor horse to follow him. Still holding the bridle, Catlin pushed his way for a mile through the tall grass, that often closed in above his head, Charley patiently swimming all the while in the thick muddy water. At length it was clear he could not keep up much longer, and his master, nearly as exhausted as himself, was just about to drop the bridle, and leave him to his fate, when they came to a spot where the banks had been trodden down by a herd of buffaloes, and, trembling with fatigue and fear, Charley staggered out, and lay down in the soft grass to be dried by the sun.

When the two travellers reached the Osage river they found it so swollen by the heavy rains that it had spread to a width of sixty or eighty yards, and had a fierce and rapid current that swept everything before it. Catlin at once unloaded Charley and tied him safely up to feed, while he wandered along the banks for some distance collecting all the wood that had been carried down by the stream and had stuck along the edge. With this he made a small raft, and on the raft he lashed his skins, his guns, his portmanteau, and even the clothes he had on. This done, Charley was driven into the river, and left to cross by himself, which he managed very well, in spite of the current, and soon might be seen enjoying his dinner on the opposite shore. Then his master plunged in after him, and pushing the raft in front, landed it safely about half a mile below. This sounds easy enough in the telling, but any one who has ever watched a river in flood, and knows the great trees and big animals that it hurries along, will understand how many things Catlin had to dodge in that short distance, and how glad he must have felt to be on Charley's back again.

FAIRY RINGS; AND THE FAIRIES WHO MAKE THEM

TRAVELLERS along the great grassy plains that extend for hundreds of miles on the eastern side of the Rocky Mountains, to an immense distance north and south, have been surprised to see dotted about, circles of grass much greener and richer than the rest of the country, to which they have given the name of 'fairy rings.'

They might watch for many moonlight nights without seeing the fairies making these rings in their dances, or if any traveller *did* happen to pass them by, he would not for one moment guess whose fairy feet tread made the grass so thick and bright. For the real fairies are huge clumsy creatures, with such enormous heads that it seems as if the animal must always be tumbling over from the weight, and they are covered from head to foot with very dark thick hair, which forms a dense shaggy mane over the head and shoulders. And these fairies have horns, not long, but very powerful, and are fond of fighting, which they have plenty of time for, as they are very sociable, and live together in large herds, sometimes, as many as three or four thousand at once. And they are known in the countries they inhabit by the name of bisons or buffaloes.

Now, it may be thought that the buffaloes which live on the prairies or plains on the borders of Texas and Mexico would lead a much easier and pleasanter life than their brothers far away up in Canada, but this is not so.

After passing a winter amidst the deep snow of the north they are a great deal fatter and stronger than if they had been spending it in the sunny south. For the grass and juicy plants on which the buffaloes feed have been completely dried up during a long hot autumn, while in Canada the animals can generally manage to get good grass by scraping away the snow, which has kept the herbage wholesome and fresh under its white covering.

Still, in Canada, as well as in Texas, the summers are very hot, and the poor buffaloes, in their thick coats, suffer a great deal. So they seek out some low plain, where they know by experience that there is a chance of finding a marshy place left by the winter's rains. What happiness for the poor tired creatures who have been walking perhaps for hours under the burning sun, carrying their huge bodies, which often weigh as much as 2,000 lb., to come on one of these little oases, as they would be called in Arabia or the Sahara! But even the prospect of a cool bath does not affect their good manners and sense of discipline. They all hold back and let the leader of the herd come forward. He sinks carefully on one knee in the soft green place, and putting down his head, stirs up the wet earth, so that the water gradually bubbles up, and a little pool is formed, though, to be sure, it is more liquid mud than anything else. When the bull has got all the water he thinks he is likely to have, he throws himself on his side, and turns himself round two or three times till he has made a circular pond large enough to cover him almost entirely. Then he feels comfortable again, and comes out, such a mass of mud from head to foot that it is wonderful how he manages to walk at all.

Sometimes it happens that the leader of the herd will not take the trouble to make the pool for himself, but suffers one of the other bulls to begin the work, which is the most difficult part. However, when this is done, the leader (who has been chosen by the rest as being stronger and a better fighter than any of the other bulls) comes forward, and goes straight into the bath that has been prepared for him. The remainder of the herd stand humbly by till he has had enough, and the moment he steps out, the bull of next importance steps in, and so on till all have had their turn. By this time the hole is often fifteen or twenty feet across, and two feet deep, and into this the water gradually bubbles up. In a few years' time the place is covered with fresh green grass, that looks even greener by the contrast with the burnt-up stuff that surrounds it.

And this is how the fairy rings are made!

Perhaps the finest of all the bisons or buffaloes are those which inhabit the country now known as Dakota, where, fifty or sixty years ago, dwelt the Sioux Indians, a nation of mighty hunters. The animals are very useful for many purposes, and while their skins are of great value as beds or coverings, the flesh forms the principal food of the tribe. The hunting is almost always done on horseback, and the first thing necessary is to catch one of the small breed of horses which formerly roamed in bands over the prairies. This little creature—it never grows much larger than a pony—is carefully trained for some years in racing and jumping and other exercises, and in the end is able to outrun any other wild animal to be found on these western plains.

Sometimes it happens—or did, fifty or sixty years ago—that for a long time together no buffaloes will pass along a certain tract of country, and then the Indians of the district suffer from famine, and are even in danger of dying of starvation. Then what joy in the camp when a scout comes in one day with the news that a herd of buffaloes are grazing not many miles off. In a moment a hundred young 'braves' have thrown aside their shields and every other heavy thing they have about them, especially any part of their dress that might be a hindrance in running for no one knows how a buffalo hunt may end.

Armed only with a bow and five or six arrows, or else with long lances, they mount their strong and swift little horses, and dash off at full speed to the grazing ground of the buffaloes.

As the hunters draw near the herd they divide into two parties, so as to surround the animals completely. If the buffaloes were to form in line and charge the enemy their great strength and bulk might tell, and they would stand a good chance of getting through. But, instead,

HUNTING THE BISON

they lose their heads and are thrown into confusion; they tumble over each other, and cannot get up again, and the Indians close in, and, galloping past, plunge the lance or aim the arrow straight at the heart, and the buffalo falls dead where he stands.

If the Indian is hunting alone he carefully chooses some large fat bull, and manages to separate him from the rest by heading him off in the opposite direction. The horse knows quite well what its master wants, and when the buffalo is well away, gallops close to it on the right side so that, at the moment of passing, the rider can turn in his saddle and aim at the shoulder. Directly the horse feels that his master has had time to give the death-blow he sheers off at once, without giving a chance for a second shot, for horses are very nervous and timid creatures, and have a very keen sense of possible danger.

If the hunters are many, and the herd a large one, there are sure to be a number of accidents both to men and horses, and indeed the thick dust often makes it difficult to see clearly till it is too late. Often, too, both men and horses get so excited that they forget their prudence, and at last have to fling themselves from their horses and trust to their own legs, or save themselves only by tearing off the buffalo skin which forms a waist-belt, and dashing it over the eyes of the buffalo.

When a great hunt of this kind is over—and it is wonderful how short a time it lasts—the Indians lead their horses through the battlefield, drawing out their arrows from their dead prey, and seeing by the private marks on the arrows themselves how much of the spoil belongs to each man. This business settled, a council is called, and the hunters seat themselves in a ring on the ground, smoking their long, gaily decorated pipes. Then, men and horses having had a rest, they ride quietly back to the encampment.

The first thing to be done on reaching the village is to choose out some of the braves to inform the chief of the success of the expedition—how many buffaloes have been killed, and how many horses or men have been lost. Next, all the women and children are sent off to bring back the meat, and a hard task it is, for they have to skin the animals and cut them up, besides carrying them home, and it seems as if the weaker ones might die on the way.

In the winter, when the Indian is in need of meat, he has to trust to his own cunning to get it, for in the colder parts of the country the horse cannot be used at all for hunting. So out he goes on his snow-shoes, which prevent his sinking into the drifts, piled up by the wind to a great depth in the hollow places. The huge buffalo, which has no snow-shoes, comes thoughtlessly down from feeding on the grass tracts which the wind has blown bare, and flounders straight in. Once there he cannot get out again, and the Indian comes up and plunges his lance right into his heart, so that he is dead in a moment. Then his skin, always in its best condition during the winter, is sold to traders in fur, and the parts of the flesh which the hunter does not want, or cannot carry away, are left to the wolves.[1]

[1] All this was true years ago. Now, for want of Game Laws, buffaloes are nearly extinct.

HOW THE REINDEER LIVE

There is perhaps no animal in the world so useful to man as the reindeer, at least none that can be put to so many uses. The flesh of a sheep is eaten, and its wool is woven into cloth; but then we should never think of harnessing a sheep even to a baby-carriage. A camel serves, in the desert, the purpose of a van and of a riding horse in one, and his hair makes warm and light garments; but he would give us a very nasty dinner, and the same may be said of some other useful creatures. A reindeer, however, is good to eat, and makes an excellent steed; its milk is nourishing; the softer parts of its horns, when properly prepared, are considered a delicacy; the bones are turned to account as tools; the sinews are twisted into thread, and, all the long winter, the skin and hair keep the dwellers in the far north snug and warm. Take away the reindeer, and the inhabitants of every country north of latitude 60°—sometimes even south of it—would feel as helpless as we should in England if there were no more sheep or cows!

Reindeers live, by choice, on the slopes of mountains, and require no better food than the moss, or little Alpine plants, which they find growing in the crevices of the rock. Sometimes, in very cold places, or when the winter is particularly severe, they take shelter in the forests; but when spring is in the air once more, out they come in great herds, thin and sore from the bites of newly awakened insects, and wander away in search of fresher pasture. In August and September, when the sun has grown too strong for them, they seek the shade of the woods again.

In their wild state reindeer are great travellers, and as they are very strong, and excellent swimmers, they go immense distances, especially the reindeer of North America, who will cross the ice to Greenland in the early part of the year, and stay there till the end of October, when they come back to their old quarters. They are most sociable creatures, and are never happy unless they have three or four hundred companions, while herds of a thousand have sometimes been counted. The females and calves always are placed in front, and the big bucks bring up the rear, to see that nobody falls out of the ranks from weakness.

We are accustomed to think of a reindeer as having thick brownish hair, but this is only partly true of him. Like many animals that live in the north, the colour of the hair is different in winter from what it is in summer. Twice a year the reindeer changes his coat, and the immense thick covering which has been so comfortable all through the fierce cold, begins to fall in early spring and a short hair to take its place, so that by the time summer comes, he is nice and cool, and looks quite another creature from what he did in the winter. As the days shorten and grow frosty, the coat becomes longer and closer, and by the time the first snow falls the deer is quite prepared to meet it.

Though reindeer prefer mountain-sides when they can get them, their broad and wide-cleft hoofs are well adapted for the lowlands of the north of Europe and of America, which are a morass in summer and a snow-field in winter. Here are to be seen whole herds of them, either walking with a regular rapid step, or else going at a quick trot; but in either case always making a peculiar crackling noise with their feet. They have an extraordinarily acute sense of smell, and will detect a man at a distance of five or six hundred paces, and as their eyes are as good as their ears, the huntsman has much ado to get up to them. They are dainty in their food, choosing out only the most delicate of the Alpine plants, and their skins cannot be as tough as they look, for they are very sensitive to the bites of mosquitoes, gnats, and particularly of midges. Reindeer are very cautious, as many hunters have found to their cost, and mistrustful of men; but are ready to be friendly with any cows or horses they may come across, which must make the task of taming them a great deal easier. They have their regular hours for meals too, and early in the mornings and late in the evenings may be seen going out for their breakfasts and suppers, which, in summer, consist, in the highlands, of the leaves and flowers of the snow-ranunculus, reindeer sorrel, a favourite kind of grass, and, better than all, the young shoots of the dwarf birch. In the afternoons they lie down and rest, and choose for their place of repose a patch of snow, or a glacier if one is at hand.

In order to tame a reindeer, you must catch him when he is very young, and even then it is no use to expect him to become as friendly as a cow or a horse. He always has something half wild about him, which peeps out every now and then when you least expect it, and often when it is extremely inconvenient. The tame reindeer is his master's pride and stay, his joy and his riches, and often his torment too! A Laplander who owns a herd of a few hundred reindeer thinks himself the happiest man on earth, and would not change lots with anybody. Yet, after all, it almost seems as if he belonged to the reindeer, and not that the reindeer belonged to him! Where they choose to go, he must follow, and neither marshy ground, nor seas, nor rivers, nor anything else, make any difference to them. For months he spends his life in the open air, bitten by insects all the summer, and by frost all the winter, for he continually finds himself in places where no wood is to be got, so he cannot have even the comfort of a fire. Food and water are not always to be had either, and sometimes, in the end, he becomes almost as much a wild animal as the reindeer themselves. When he eats, he eats strange things; as for washing, he never thinks about that at all. His sole companion is his dog, with whom he shares whatever he has; but all his hardships seem light, for are they not suffered for his beloved herd?

In Norway and Lapland great herds of reindeer may be seen, during the summer, wandering along the banks of rivers, or making for the mountains, returning with the approach of winter to their old quarters. With the first snow-fall they are safe under shelter, for this is the time when wolves are most to be feared. In the spring they are let loose again, and are driven carefully to some spot which is freer from midges than the rest. And so life goes on from year to year.

Reindeer herding is by no means so easy as it looks, and it would be quite impossible, even to a Lapp, if it were not for the help of dogs, who are part of the family. They are small creatures, hardly so big as a Spitz, and very thin, with close compact hair all over their bodies. These dogs are very obedient, and understand every movement of their master's eyelid. They will not only keep the herd together on land, but follow them into a river, or across an arm of the sea. It is they who rescue the weaklings in danger of drowning, after their winter's fast, and in the autumn, when the reindeer have grown strong from good living, drive the herd back again through the bay.

A herd of reindeer on the march is a beautiful sight to see. They go quickly along, faster than any other domestic animal, and are kept together by the herdsman and his dogs, who are untiring in their efforts to bring up stragglers.

When a good stretch of pasture is found, the Lapps build a fold, into which the reindeer are driven every evening, so that the work of the milkers may be lightened. These folds are made of the stems of birches placed close together and strengthened with cross pieces and strong props. They are about seven feet high, and have two wide doors. At milking-time, which the dogs know as well as the men, the animals are driven inside by their faithful guardians, and milking begins busily. The young ones are generally left outside to graze or play, under the watchful eyes of the dogs, who see that they do not wander too far away. Inside the fold the noise is really deafening. The reindeer run to and fro, giving loud cries and throwing their heads about; which, as their horns are very big, is not pleasant for the milkers. Any one walking that way would be struck, first, with the sound of the movement and commotion going on in the enclosure, and this would most likely be followed in a few minutes by a crackling noise, as if a hundred electric batteries were at work at once.

In the middle of the fold are thick tree trunks to which the reindeer which have to be milked are fastened, for without these they would not stand still one single instant. The milkers have a thong which can be thrown round the neck or over the horns of the animal, and this is drawn closer till it is tied by a slip noose over the creature's mouth, so as to prevent it from biting. Then the ends are made secure to the milking block, and the milking begins at last—the animal all the while struggling hard to get free. But the Lapps know how to manage them, and only draw the cord tighter over the nose, so that the creatures are bound in self-defence to remain quiet. The milk flows into a sort of large bowl with handles, but the Lapps are both careless and dirty in their ways, and not only waste a great deal of the milk, but leave so many hairs in it that it is necessary to strain it through a cloth before it can be drunk. However, the milk itself is very good, and as thick as cream, and makes excellent cheese. The milking once over, the doors are opened, and the animals scamper out joyously.

Still, altogether, the life of the owner of a herd of reindeer cannot be said to be an idle one. He is always on the tramp, always on the watch ; he suffers thirst and hunger, cold and fatigue, and it is very lucky he is in general so well satisfied with his lot, and thinks himself the most fortunate man in the world.

THE COW AND THE CROCODILE

Crocodiles are found in nearly every large river all over the tropics; they are of immense length, sometimes reaching as much as twenty feet and upwards, and are covered with a thick, scaly hide which renders them almost invulnerable. Not only is their throat very large, but it is capable of expansion, so that a crocodile can with ease swallow a small person or animal whole, though, in the case of a larger victim, its snapping jaws and immense teeth can bite through a human bone, or any equally hard obstacle, as clean and sharp as though it had been cut with a knife. These huge teeth are sixty-eight in number, thirty-four in each jaw. They are very long and sharp, and those of the upper and lower jaws interlock, so that woe betide any person seized upon by them; there is no possibility of escape, or, if by good fortune he be rescued, he will certainly leave a limb behind him in the jaws of the devourer.

It is a mistake to suppose, as many persons do, that the crocodile immediately consumes its victim; in the case of small animals, such as dogs and fawns, this may be so. Large animals, however, when seized, are dragged beneath the surface of the water, held there till drowned, then borne off to some favourite hiding-place, there to be eaten at leisure. The fore-feet of the crocodile are shaped much like a short human hand, armed in place of fingers with five long horny claws, which hold the prey whilst tearing it with the teeth.

The time when it is most dangerous to enter the water on account of these greedy monsters is towards sunset, for then the fish come to the shallow water to feed, and the crocodiles come to prey on them; they may be seen dashing furiously like huge pike after the larger fish, who often leap several feet out of the water in the vain hope of evading their pursuer.

Their cunning is only equalled by their ferocity, and nothing daunts them, not even the sight of a large steamer passing quickly through the water, from the deck of which they will even snatch any person heedless enough to place himself within their reach. This happened more than once on Sir Samuel Baker's explorations of the White Nile. A sailor, seated on deck dangling his feet over the side of the vessel within half a yard of the water, was seized and carried off so swiftly, that, though a hundred men were present, nothing more was ever seen or heard of him. Another sailor, who was seated on the rudder washing himself, was borne off just as suddenly in the sight of all his comrades.

The troops were in the habit of bathing in a small dock, which had been made for the accommodation of one of the steamers, and was connected with the river by a canal thirty yards long and only three feet deep. This was considered a perfectly safe bathing place, and free from the intrusions of crocodiles. One evening, however, the captain was absent from muster, and as it was known that he had gone to bathe at this basin, search was immediately made there for him. His clothes and red fez alone being found on the bank, a number of men went into the water in search of his body, which was not long in being discovered. One leg being broken in several places proved unmistakably that it was the work of a crocodile, who would doubtless soon have returned to devour his victim. Some months after this catastrophe another occurred in the same canal, occasioned, it was supposed, by the same monster, though there were no actual proofs of the fact. As Sir Samuel and Lady Baker were sitting out of doors enjoying the comparative coolness of the evening, a man rushed frantically past the sentries throwing himself on the ground at Sir Samuel's feet, grasped him by the legs. As soon as he could find breath he gasped out, 'Saïd! Saïd is gone! taken from my side this moment. We were wading together across the canal by the dock where Reis Mahomet was killed, when a crocodile rushed like a steamer from the river, seized Saïd, and went off with him.' Assistance was quickly on the spot, but all trace of the unhappy Saïd had completely disappeared, and not even a ripple on the surface of the water bore witness to the melancholy fact.

Another man belonging to the same expedition was less unfortunate. While gathering watercress he had his arm bitten off, and was only saved from utter destruction by his comrades holding tightly on to him.

Yet another man was seized by the leg while helping to push a vessel off a sand bank. He, too, was saved by the help of the soldiers engaged on the same work, but with the loss of his leg.

From this formidable description, a tug-of-war between a crocodile and a cow would seem a very unequal contest, and certain to go in favour of the crocodile. But on the only occasion that such a thing is known to have taken place, the cow came off with flying colours. She was one of three large cows, with immense powerful horns, brought by Sir Samuel Baker to Gondokoro, on the White Nile. Being different from, and much handsomer than, the small, active, cattle of that district, they were looked upon with great admiration by the natives. When Sir Samuel was obliged to depart into the interior of Africa, he entrusted the three cows to the care of a neighbouring chief, who, while responsible for their safety, enjoyed the use of the milk. Upon Sir Samuel's return to Gondokoro, after an absence of two years, he found not only that the cows were in good health, but that one of them had become an object of great veneration to the tribes. Every morning her horns were wreathed with fresh flowers, and she had become the sheik or chieftainess of all the herds, for she had performed the remarkable feat of having caught a crocodile.

It had happened in this way: she had gone to drink at the river, at a place where the banks sloped gradually down to the water's edge. While she drank, a large crocodile came out and seized her by the nose, with the intention of dragging her down to the water, and there drowning her, according to crocodile custom. Far from this, however, for once he found that he had met his match. The cow being heavy and strong, and the slope of the bank gradual, she succeeded in dragging the crocodile out of the water, and as the creature would not let go its hold, and the cow was equally determined and more powerful, they gradually receded several yards from the water's edge. The natives attracted by the bellowing of the cow, rushed to the rescue, and soon put an end to the combat by despatching the crocodile with their spears. Its head was kept as a trophy, and the cow became a heroine for life.

A bullock on another occasion was less fortunate, or, perhaps, less plucky and determined; a crocodile having succeeded in dragging it into the water, several times, in its struggles, its body was seen to appear above the surface, its head being held down by its captor. At length nothing was visible but its tail, writhing and twisting convulsively in the air, like a snake, till at length that too ceased to move, and disappeared. Presently the dead body rose to the surface, and was seen to float, while the triumphant crocodile swam alongside, contemplating its victim with satisfaction.

PRINTED BY
SPOTTISWOODE AND CO., NEW-STREET SQUARE
LONDON

www.ingramcontent.com/pod-product-compliance
Lightning Source LLC
Chambersburg PA
CBHW032018220426
43664CB00006B/295